The Changing Face of New England

Books by Betty Flanders Thomson

THE CHANGING FACE OF NEW ENGLAND

THE SHAPING OF AMERICA'S HEARTLAND

The Changing Face of New England

BETTY FLANDERS THOMSON

Illustrated with Maps

HOUGHTON MIFFLIN COMPANY
BOSTON · 1977

A portion of this book has appeared in AMERICAN HERITAGE.

Library of Congress Cataloging in Publication Data
Thomson, Betty Flanders, date
The changing face of New England.

Includes index.
1. Natural history—New England. 2. Physical geography—New England. I. Title.
QH104.5.N4T48 1977 500.9'74 77-4476
ISBN 0-395-25725-5 pbk.

Printed in the United States of America

V 10 9 8 7 6 5 4 3 2 1

For my mother, companion on explorations of back roads and another lover of New England

Foreword

WHEN I WAS A CHILD in northern Ohio, the packages of "construction paper" we used to cut things out of always had a few sheets of what the teacher called "red-violet." This seemed to me a peculiar name for a peculiar color that had no counterpart in the real world I knew. But years later when I went away to college in New England, I found that the hills on a clear winter afternoon took on exactly that violet shade that had so disturbed my childhood sense of reality. Moreover, here was a land where winter sunshine was bright and sparkling, and the coldest days were the clearest, ending with pink and palest yellow skies fading into amethyst and indigo shadows. So began my enchantment with the New England countryside.

Gradually it came to seem natural to live in a place that offered such delights as hill-slopes covered ankledeep with the little stars of hairycap moss; rough pastures strewn in spring with bluets like nothing so much as spilled talcum powder; brooks that tumble over round, hard boulders of granite and little shallows of clean sand; and, floating on the air of winter twilights, the spicy, delicious fragrance of white birch logs burning on an open hearth.

Then a few years ago I achieved a car of my own, and for the first time was free to wander the country roads to my heart's content. Soon I became very curious about the great expanses of rather scraggly-looking, youngish woods that nobody seemed to care much about. This was all very different from the small farm woodlots widely scattered over the highly cultivated landscape of the Middle West where I grew up.

Hunting for answers to the puzzle became more and more absorbing. The most enlightening information turned up in technical journals of botany, geology, and geography. And while much of the fascinating story is heavily interlarded with history and can readily be understood by any reasonably bright and observant person, it is no more accessible to most non-professionals than if it were recorded in Sanscrit or early Sumerian. Everyone has at one time or another said, "There ought to be a book . . ." So this is my attempt to set down for all lovers of New England how her history —both human history and natural history, for they are closely intertwined—can be read from her physical countenance just as if from a printed page, if only one knows what the signs mean.

The face of New England is somehow familiar and homelike to everyone who has ever been a schoolchild in America or who has read our literary classics. We know about Plymouth Rock and Bunker Hill, Cape Cod sand and the rockbound coast of Maine. But how much more the landscape means when the mind's eye sees Plymouth Rock not only as that first alighting place, but also as a far-traveled fragment dropped there thousands of years ago by a melting glacier; sees Bunker Hill as a little mound moulded by the moving underside of the same glacier, and Cape Cod as a ridge of sandy debris that accumulated in festoons around the melting edge of one of the glacier's forward lobes. The informed eye sees in the rocky Maine coast not only a stretch of rugged and picturesque scenery, but also the drowning remnants of ancient mountains that have been raised and lowered by movements of the very crust of the earth, and scraped and rounded by the passing of a mass of ice the size of a continent.

New England has had a long history, not only in relation to the nation of which she is a part, but also in relation to the history of the planet. The folded and faulted rocks that form her bony structure are so ancient that their exact history has not yet been fully deciphered; but most of us know them at least vaguely as examples of "old-worn-down-mountains" in contrast to the "young-rugged-mountains" of the West. The marks of the last wave of glacial ice, on the other hand, are clear and fresh, and lie about us everywhere. Once one learns to see them, the glacier seems a very real and tangible thing, and the twelve thousand years since the ice disappeared become as the twinkling of an eye.

The vegetation that clothes the land is just as pithy with meaning. The forests that once covered all but a few corners of New England have come back from the south, seedling by seedling, in the time since the ice last began to retreat. They have changed with the climate and they have changed with the lives of the men who have lived here, first red men and then white men. Signs of the changes fairly proclaim themselves in every pinewood whose understory of young growth consists of broadleaved trees, in every fire-begotten birch grove and every stand of stump-sprout coppice.

New England's story is fully as dramatic as that of Niagara Falls or the Grand Canyon, and as readily comprehended. There is far more in it of strongly human interest than there is in those vast geological museums. So here it is, set down in these pages so that having eyes, one may both see and understand.

Contents

The Changing Face of New England

I

Forming the Framework

NEW ENGLAND is one of the oldest continuously surviving lands on the face of the earth. It was far back in the mists of Cambrian times, half a billion years ago, that the ancestors of our Berkshires and Green Mountains emerged from the waters of an ancient inland sea. Even as geological time is reckoned, that is a long time ago. And while the ocean waters have advanced and receded, and hills have risen and worn away again, always there has remained in New England some area of dry land exposed to wind and weather.

Through all these aeons, running water has been slowly and steadily wearing away at the land, moving little earth particles from the higher places to the lower. Bit by bit the upland rock was torn apart by the abrupt violence of sudden freezing and thawing, or ground to powder by the slow scraping of cobble on boulder in the beds of mountain streams; and year by year and grain by grain, the ceaselessly running water slowly shifted the weight that lay upon the earth.

In time the earth responded to the changing balance of its burden, and all of what was to become New England slowly buckled into great north-and-south-running folds. When in time the folded rocks could bend no more, a gigantic cracking and tearing of the very fabric of the earth released the accumulated tensions. As the mighty forces continued to press, the broken rocks were thrust high in the air to form mountains that must once have been far more spectacular than the remnants that have survived the wear and tear of three hundred million years to the present time. Even today the underlying texture of our landscape reflects the pattern of those ancient ripples and convulsions.

While this was going on, farther to the east the ancestors of the White Mountains were born in the upwelling of a vast, dome-shaped mass of molten granite from deep within the earth. The intruding rock probably never reached the surface; but it must surely have thrust the level land above it high into a mountainous bulge.

Even as the mountains were rising, the slow, persistent wear of running water began the work of leveling. For aeons, sediments washing down from the heights accumulated layer upon layer on the lower slopes and in the central lowland that even then existed as the remote ancestor of our Connecticut Valley. Dinosaurs in their day wandered over the alluvial mud, leaving their footprints on the rippled flat. As time went on and rivers kept flowing, the muddy flats became buried beneath still more sediments until eventually the whole mass was compacted into rock by the weight of the overlying burden. Meanwhile from time to time lava issued from volcanoes or from cracks in the earth and flowed out in sheets over the land surface. In time the layers of harder volcanic rock became interleaved with the softer erosional sediments like the frosting in a giant ribbon cake.

Still the waters ran, and again the earth's crust stirred under its shifting burden. One of the mighty stirrings set off a hundred-mile-long crack that tore along the eastern edge of the primordial valley. Along the western edge of the rift, the valley floor gradually settled downward for hundreds, even thousands, of feet, leaving a sharp depression some twenty miles wide across the center of present-day Connecticut and Massachusetts. Its descendant remains as a lowland reaching from Greenfield to New Haven, and traversed as far as Middletown by the Connecticut River.

At last the land lay quiet for so long that nearly all New England was reduced almost to sea level and as nearly to a flat plain as flowing water could wear it. Above the great plain rose only a few scattered stumps of mountains. These survived to the end because they contained more resistant rocks, or because they chanced to form the local watershed, the last of any upland to give way before the waters. The north was less completely leveled, and there the mountains persisted in groups or ranges. The southern outliers, like Mount Monadnock in southern New Hampshire, are more isolated. This ancient peak rears up above the surrounding general skyline so clearly and so characteristically that its name has become a generic term for any such isolated remnant mountain wherever it occurs around the world.

Later, toward the end of Cretaceous time, the eastern margin of the smoothed-off plain sagged a little, letting the sea wash in over it. In the shallows of this swampy flatland enough vegetation accumulated among the river-borne sediments to form a little low-grade coal, which erosion has since uncovered again in the Boston Basin and around Narragansett Bay.

It was then, when our eastern mountains were already old and worn, that far to the west the land was lifted up from the bottom of the sea and thrust miles into the air to form the Rocky Mountains.

Then once more the crust of the earth stirred under all New England, and once more the land rose, but this time so gently that the surface rocks were not disarranged, nor the monadnocks disturbed. With the old flatland tilted hundreds of feet above the sea, the sluggish rivers of the plains took on a fresh vigor that set them off to a new flush of valley-carving.

The newly formed valleys had just begun to become broad and smooth when once more a gentle uplift took place, and the reinvigorated streams began once more to cut steep new valleys on the broad old valley floors. The soft sedimentary rocks of the central lowland were rapidly worn away almost to sea level. Only the harder lava sheets that lay buried among the sediments resisted the forces of erosion. Their tilted edges remain today as conspicuous traprock ridges strung along the wide valley floor from New Haven to Greenfield. The harder rocks athwart the valley eroded even more slowly and still form a rolling upland of hills and ridges over most of New England.

The higher New England hills have their counterpart in the Adirondack Mountains of New York State. Here, too, an ancient group of hard-rock mountains became worn down into a gently rolling landscape, with the highest peaks along the watershed left standing as monadnocks that rose above the general skyline. Then later, the whole surface of the land was lifted up repeatedly so that new cycles of erosion carved it with a series of new valleys.

The broad Connecticut Valley with its traprock ridges is matched in age and history by the lowland that stretches southwestward from New York City across New Jersey and beyond. Here, too, a region of soft, sedimentary rocks interleaved with harder lava sheets was worn to a sea-level plain, lifted up and once more eroded. In the present state the soft rocks are down to sea level, with marshes covering large expanses of the land, and the harder rocks still stand as ridges that we know as the Watchung Mountains in New Jersey and the Palisades along the Hudson River.

In the New England upland the newest generation of valleys was still young and steep when the whole story was interrupted by the coming of the great Ice Age.

It began gradually about a million years ago. Almost imperceptibly, all over the world the weather became a little cooler and a little damper. The change need not have been great, just a drop of three or four degrees in the average year-round temperature, a few more wet days each year. Wherever the land stood high and cold, as it does in northern Quebec and Labrador, a little more of the year's moisture fell as snow, and the drifts lay a little longer in the spring. Soon the snow began to accumulate faster than it melted away each summer. For decades and centuries it piled up, until its growing weight gradually compressed the lower layers into ice. Finally the accumulating mass became so vast and so heavy that it squeezed down and outward of its own weight, and began to flow. For thousands of years the ice cap grew, oozing out in all directions from its northern center, covering at last all of New England as well as much of the Middle West and even the northernmost part of the Great Plains. In time the ice became so thick that even the highest mountain tops were completely washed over.

As the front of the ice crept down over the land it swept away all things movable—loose soil and little plants, great rocks and entire forests. Here and there the ice froze onto protruding hummocks or ledges of bedrock and then, moving on, plucked out great chunks of solid earth to form such craggy sculptures as the Old Man of the Mountain in Franconia Notch, New Hampshire.

Some of the rocks gathered up by the moving ice were carried hundreds of miles and eventually discarded as "glacial erratics" far from the nearest outcrop of matching rock. Many more boulders were carried only a few feet or at most a few miles and then dropped at random, sometimes perched improbably on a bald hilltop, or sometimes strewn out in a train of broken fragments. Such boulder trains have been traced for miles from certain outcrops of characteristic rock. Some of the largest of them are found in Rhode Island and fanning out southward and southeastward from Mount Ascutney, Vermont, and Red Hill in New Hampshire. Much smaller boulder fields, thickly strewn with rocks of all sizes, are common almost anywhere.

That the ice moved generally southward over New England we know from the trends of long boulder trains and from the direction of grooves and scratches ground into exposed bedrock. Even the hard rocks of New England yielded to the grinding and scraping of the glacier's massive weight. Bulges on the upstream sides of protruding hills and ledges were pared off, while downstream sides were steepened by "plucking" and "chattering" of the ice. Surfaces that had been polished smooth became streaked with grooves and scratches cut without reference to the grain of the bedrock by smaller rocks frozen into the underside of the moving ice. Many such scars must since have become hidden by debris; but there has not been time enough since the disappearance of the ice for the rock ribs of this region to be weathered very much, and many grooves and scratches can still be found scored across flat or gently rounded surfaces that are not yet covered over with plants and soil.

Sometimes a rocky knob protruding into the bottom of the moving ice caught at the stickier parts of the debris that had been worked into the sole of the glacier. Clay sticks to clay more tenaciously than it does to ice, so that the clayey parts of the passing glacier's load accumulated around the first little nuclei and eventually became formed into low, streamlined hills that we know as drumlins.* A drumlin is shaped much like the inverted bowl of a spoon, with the handle end pointing upstream and the shallower, tapered tip trailing off down the current. Most of them are a mile or so long and of the order of a hundred feet high. They commonly come not singly, but in hundreds. South-central New England has some three thousand; but there are five thousand in southeastern Wisconsin and ten thousand in central and western New York State!

In the hollows and in steep valleys lying crosswise to the direction of the glacier's flow, stagnant ice seems sometimes to have protected the land surface from the glacial bulldozer, and such places lack the almost universal marks of the ice's passing.

Four times the ice came down and covered all of present New England, and four times it has melted back far to the north. Each ice wave obscured most of the effects of its predecessors; but enough signs remain of the earlier times to piece together the probable story.

It appears that the intervals between the floods of ice were probably even longer than the glacial periods. Fossil plants dating apparently from the last interglacial period have been found in rocks near Toronto. This means that after the ice had once disappeared from that area, enough time passed for plants to return from the south and become buried and changed into fossils before the ice came back.

* *Drumlin,* an Irish name for such hills, comes from the Gaelic word *druim,* the crest of a hill.

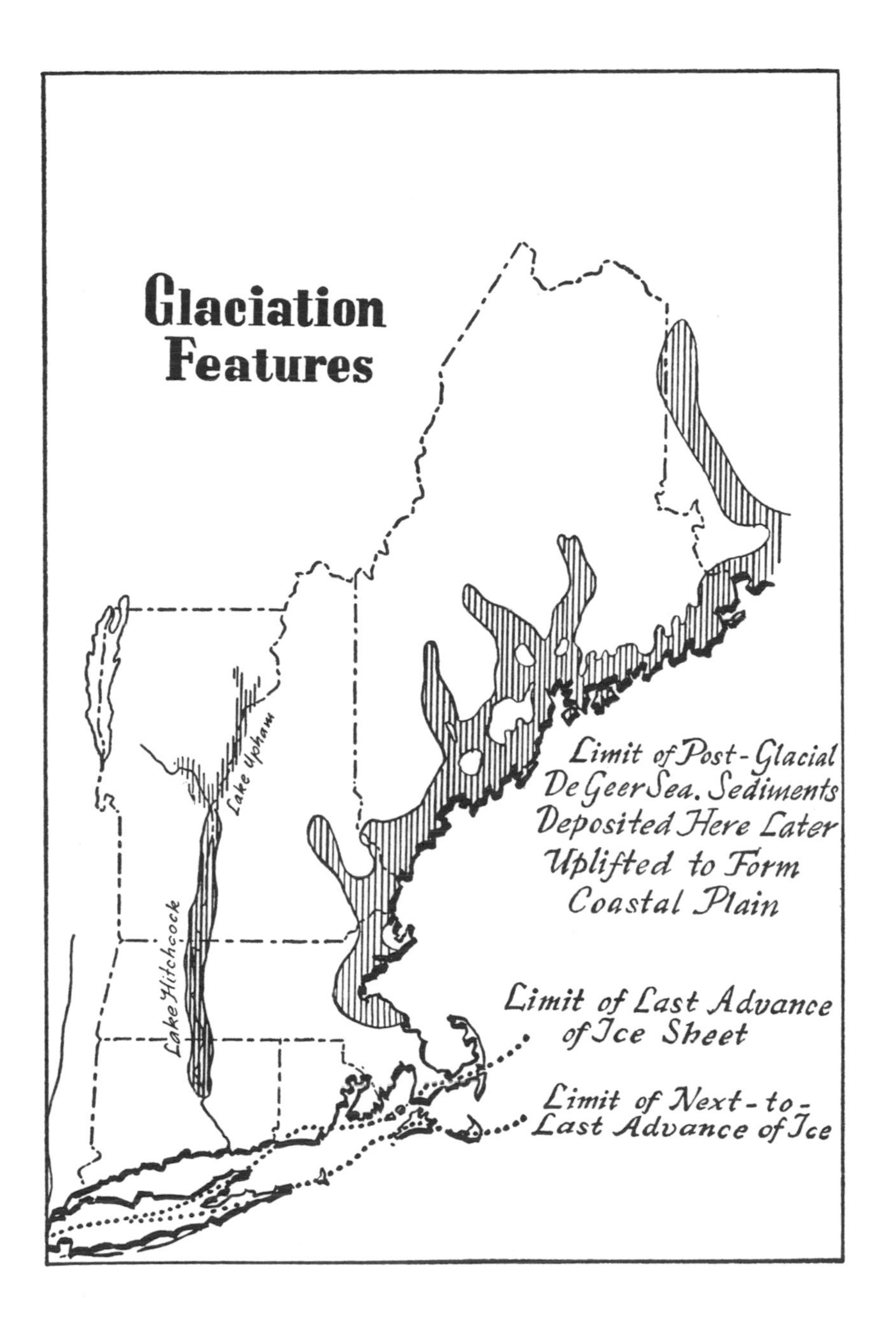

Glaciation
Features
Lake Upham
Lake Hitchcock
Limit of Post-Glacial
DeGeer Sea. Sediments
Deposited Here Later
Uplifted to Form
Coastal Plain
Limit of Last Advance
of Ice Sheet
Limit of Next-to-
Last Advance of Ice

Each time the ice came down it continued to move as a living flood for thousands of years. Through long periods of time replenishment from the north was just balanced by melting away at the south, so that the southern edge of the ice sheet was held stationary. Along its outer fringes a jumble of rocks and gravel accumulated just where they fell out of the melting ice, forming a long ridge or terminal moraine. Torrents of meltwater, issuing from below and above and even from tunnels through the glacier, flowed out over the top of the growing moraine. Then as the water flowed down off the moraine onto level land or into the edge of the sea, it spread its sediments in broad outwash sheets and fans that sloped gently out beyond the morainal ridge.

One of the ice sheets left its morainal high-ice mark along the northern shore of Long Island, stretching from Harbor Hill east to Orient Point, where it dips below the present water level to appear again as Fisher's Island, part of the Rhode Island shore, and the Elizabeth Islands and Cape Cod. An earlier ice sheet reached a little farther south, to the middle of Long Island, Montauk Point, and Martha's Vineyard and Nantucket. The records of others perhaps lie submerged offshore beneath the ocean.

Farther west the southern edge of the ice sheet can be traced across northern New Jersey. Then it loops up around the northern end of the Alleghanies into New York State and follows roughly down the Ohio River, up the Missouri, and north across Canada to the Arctic.

It was some fifteen thousand years ago that, for reasons not yet clearly understood, the weather moderated so that for the last time the ice ceased to advance and, in so stagnating, ceased to be a glacier. As melting quickened, the edge of the stalled ice sheet was eaten away northward. At the same time the thawing mass wasted from the top downward, so that the highest land was cleared of ice first, and hilltops appeared one by one like islands rising in the sea of rotting ice.

As the great thaw progressed and the higher lands once more stood naked to the winds and rains, water streamed down over every slope of the rough topography. For a long time valley bottoms remained filled with masses of stagnant ice. These often blocked the normal drainage outlets, so that many long, narrow lakes formed in the grooves along the contact lines where the old ice lay against the hillsides. Wherever ice or a mass of ice-gathered debris blocked the old drainage channel, water became ponded behind it, leaving the overflow to escape over a high-level spillway however it could. Gradually as the ice subsided, new and lower spillways would be uncovered, so that the water level of the lake dropped abruptly, perhaps in a series of steps.

The traveler over the rough New England landscape often is aware of a change of scene as he emerges from the hills onto a level tract of relatively stone-free soil. Sometimes these areas are dignified by a local name such as "North Plain" or "Fog Plain," or even "Great Plain." These mark the sites of just such glacial lakes, where incoming streams, suddenly checked by the drag of quiet water, laid down their burdens of silt and sand. Gradually the lake bottom became spread with a smooth, level floor that eventually made one more patch of plowable land tucked among the hardscrabble.

Often the old lake-bottom deposits do not completely fill the valley floor, but hang like terraces part way up the valley sides. In such cases the glacial lake lay in a temporary valley that had one of its sides formed by the upper hillside and the other by the edge of a thick tongue of ice lying against it. When the ice melted, one side of the valley simply disappeared; and as the water level dropped, the old lake bottom was left hanging high and dry on the hillside above the new level of the lake surface. Sometimes one can trace a series of old lake-bottom terraces, descending like a flight of steps, each perhaps ten or twenty feet high, and each marking a successive stage of the falling level of an extinct lake.

Sometimes water streaming off the exposed upland swept down into ice-clogged valleys that held no standing water. Then it flowed out over the ice surface and spread its sediment over the top. After the glacier began to stagnate, much of it probably became covered over quickly in this way with rock debris that formed a crude soil where hardy pioneer vegetation could find a foothold. There is a dying glacier in Alaska today whose lower end is covered with a thick layer of just such a raw soil. On it grows a luxuriant spruce forest that merges imperceptibly into the nearby land-based forest. There is no reason to doubt that things were much the same in parts of New England while the ice was on the wane.

In the thousands of years of its travels the ice sheet had become heavily loaded with earth and rocks of all sizes. As the ice thinned and finally melted away altogether, it eased down onto the land this long-accumulated load. To this day the surface of the higher land bears a vast ground moraine, a patchy and tattered mantle of unsorted rubble that may be anything from a few inches to many feet thick. A road cut or a cellar excavation dug into such glacial "till" reveals what a hodgepodge of clay, sand and gravel, and rocks even up to house size, the ice set down on dry land.

The jumbled ground moraine is quite different from the clearly layered sediments that were laid down under water, coarse or fine as the current that dropped them ran fast or slow. Such old lake and stream beds are much exploited as good sources of sand and gravel. In many a modern gravel pit where the earth is exposed in a fresh vertical cut the story of ancient events can still be read. Sometimes the sequence of layers is so complex that only a long-practiced expert can decipher the story. But here and there one finds a place that shows a straightforward, gradual change from clay at the bottom that can settle out only from deep, quiet water standing in a lake, through silt and sand as the current moved more quickly through a shallower lake bed, and on to coarser gravel and cobbles that could have been brought in only by fast-running water flowing over the top of the filled-in lake or pond.

Within some of the finer silts and clays one can distinguish many separate layers, each beginning with relatively coarse silt on the bottom, shading gradually off finer and finer and ending with the smoothest slippery clay. Geologists believe that each of these layers represents a year's accumulation of sediment on the bottom of a lake or pond. In the spring, when rain and melting snow make every stream run full and fast, relatively large, almost sandy particles are carried well out into the middle of the lake before they settle. As spring gives way to summer, brooks subside and flow more gently into the lake, carrying with them only finer, silty sediment. At the last comes winter, when a covering of ice stops all but the slightest of underwater motion. In such stillness even the finest clay can settle to the lake bottom. The end of the cycle comes abruptly when the ice breaks up in another spring and turbulent streams again come rushing in with their silt and sand to start a new layer, or "varve," as geologists call it, marking the start of another year's cycle. Such varved deposits are a most useful tool for measuring the passage of time. Good clay beds may show hundreds of varves, representing as many years. From them the life span of many a postglacial lake has been figured.

When at last the glacial ice had melted away and water began to drain from flooded lands, the face of the earth was completely denuded of both animals and plants. Now, most animals are able to gather themselves up and move from place to place as conditions change. Newly hatched oyster larvae swim about and attach themselves where the water temperature suits them, whether northward or southward from the homes of their forebears; and woolly mammoths can easily walk to wherever the grazing looks good to a mammoth. But what about plants?

At first thought one might guess that whole broad belts of vegetation migrated southward before the advancing ice and then returned northward in its wake as it retreated. This does not appear to have been the case, however. We know that the epoch of cool, wet weather that sent the ice sheets creeping down over New England affected the entire world. Europe and Asia had their own continental ice sheets at the same time, and snow lines on mountains even in the tropics stood two to three thousand feet lower than they do now. But this was as much a matter of moisture as of temperature, and the glacial cold did not extend so far beyond the ice front as one might think.

The scattered fossils that we have from glacial times show that some northern plants such as spruce and larch did indeed live farther south then than they do at present. We have further evidence from plants living today. Scattered through a narrow strip of country that parallels the southernmost edge of the glacier there are shady cliffs and cold bogs where we find plants such as cottongrass and cranberries that normally grow in far more northern regions. In these relict islands of coolness conditions of life are much more like the north than they are like the immediately surrounding country; and here the survivors from another age are able to hold out for a while longer in a world become warm and dry. All this means that some plants did indeed move south temporarily before the advancing ice.

But there are many more kinds of plants that seem to have been little affected by the nearness of the great mass of ice. If one knew nothing of geology but studied only the geographical distribution of plants at the present time, he would be forced to believe that something has happened to the plants that grow through New York and Pennsylvania and on across the Middle West—something that destroyed many kinds of plants in the region lying north of an irregular but clearly marked line that can be traced across this region. While most species in the region grow on both sides of this line, there are a number that stop short just south of it. The key plants include mistletoe, sweet buckeye, and Hercules' club. This remarkable dividing line shows no relation to any discernible factor of modern soil or climate, but it coincides closely with the southern edge of the ice sheet as geological evidence shows it to have stood. The conclusions seem inescapable that on the one hand it was the ice that eliminated the plants from north of the glacial border, and that on the other hand the plants just beyond the edge of the ice were little affected by its proximity.

In New England every inch of the land was swept clear of vegetation by the glacier, and all our plants—or their ancestors—have moved in within less than fifteen thousand years. Since new land just cleared of ice is not a hospitable place for plants to grow, and only a few kinds, like spruce and alder, can pioneer onto raw, ground-up rock that contains no organic matter to make it a true, fertile soil, botanists have been understandably curious about how the plants came back after such a cold, clean sweep. Fortunately we have a full and legible record in the form of bog peat deposits.

In grinding over the land the glacier changed many details of topography. Very commonly the changes interfered with drainage and left catchment basins where all sorts of debris could accumulate under standing water and so be preserved from decay. Many such depressions remain today, still containing their ponds or lakes.

In the bottoms of these hollows lies an accumulation of material that contains a key to what has happened in the vicinity since the ice disappeared. Ever since the pond was born, debris of all kinds has been piling up on its floor. Spring freshets wash in soil that has eroded from the land upstream. Summer storms break off leaves and twigs that are blown into the water by gusty winds. There is driftwood of all kinds, and the autumn's fall of leaves every year, and the spring's clouds of pollen. If the pond has no surface outlet, or if water currents flowing through it are slow enough to let all this rubbish settle to the bottom, the organic matter may have reached a depth of many feet in the long years, even after partial decay has greatly reduced its bulk; and the size of the little body of open water left in the middle may have shrunk considerably. Such a situation is, of course, what we call a bog. Like the other debris left in various places by generations of prehistoric human inhabitants, bog deposits have been much studied for what they reveal about past times.

When the deposits in a deep bog are examined in detail, the lowest layers are found to consist entirely of mineral silt and clay. These would have been laid down soon after the nearby land had been cleared of ice and while vegetation was still too sparse to form a protective cover on the bare earth. Above the mineral floor lies a mass of peaty organic matter that consists mostly of plant remains. Much of this is decomposed beyond recognition; but bog deposits contain large numbers of pollen grains, which have a tough, waxy outer coat that strongly resists decay, even after thousands of years.

Every year each wind-pollinated tree and shrub and grass flower produces prodigious quantities of dustlike pollen, of which only a fraction finds its way to another flower where it can take part in the formation of seeds. The rest is wastefully carried away on the wind until it falls to the earth, or drifts onto the surface of a pond somewhere. There it may float as a golden film, remaining for days if the water is still; but eventually it filters through to the bottom, adding to the miscellany that has already accumulated there. This has gone on as long as there have been plants near the pond, and the record continues even to the present time.

It is a great boon for our knowledge of the past that the minute pollen grains of plants have characteristics as particular and almost as varied as the larger traits by which we distinguish one species from another. An experienced person can readily tell what kind of plant produced a given pollen grain. It makes no difference whether the plant is living today or lived out its life many thousands of years ago.

The analysis of pollen deposits has become a highly developed science in recent years, but its method is essentially simple. The "palynologist," or student of pollen, proceeds into his chosen bog armed with a long, slender rodlike contrivance that pierces the soft peat to any desired depth up to about forty feet. There it punches out a cylindrical core that can be pulled to the surface intact. By taking samples at successively greater depths, it is possible to obtain a complete core from top to bottom of the peat. In this core is contained a full record of the history of the bog.

Back in the laboratory, the investigator takes a series of small bits from his peat core at intervals of several inches. These he examines with his microscope and identifies the first one or two hundred pollen grains he finds among the debris in each sample, counting the numbers of each kind. When the job is finished, he has a chart known as a "pollen profile." This shows the proportion of the various kinds of pollen at each level in the peat, the lowest being the oldest; and from it can be read the sequence in which different kinds of plants came into prominence in the vicinity of that bog.

Pollen profiles from New England and Canadian bogs show that while the glacier was disappearing, such vegetation as there was consisted of tundra plants, mostly grasses and sedges and other small herbs, such as wormwood, plaintain, and fireweed. Then, as soon as the ice was all gone from any locality, the forest came back. The first trees to appear were preponderantly spruce and fir, along with a few smaller plants, such as alders, arctic willows, and some of the heaths. It may well be that the spruce and fir moved in even before the ice had all melted out from the glacial ground moraine, since they grow today close to the northern tree line where their roots occupy only a shallow layer on top of the permanently frozen subsoil.

As soil conditions and climate improved, a second group of trees moved out behind the advance guard of spruce and fir. This was a mixture of hemlock, white pine, and northern hardwoods, mostly maple, birch, and beech—the kind of forest that still clothes the greater part of the New England upland.

All of these early comers had of necessity retreated southward before the advancing glacier. While the ice age was on, they took refuge in the cooler places on the higher Alleghanies and perhaps also in the New Jersey pine barrens. Then when the tide turned and the ice began its northward retreat, the hardiest plants crept out first and followed behind it, spruce and fir in the vanguard, and then the hemlocks, birches, and maples.

As conditions moderated still further, the returning northern refugees moved on a little and gave way in southern New England to a more southerly group. These were part of a rich and varied assortment of hardwoods that has been developing in the Cumberland Mountain region of Tennessee and Virginia since long before glacial times. This proliferous forest community includes beech, tuliptree, several kinds of basswood, sugar maple, chestnut, sweet buckeye, red oak, white oak, hemlock, and silverbell. These are only the dominant trees. Birch, black cherry, cucumber tree, ash, red maple, sour gum, black walnut, and several hickories are also common; and over a dozen more species appear here and there among the more abundant kinds.

Flowering plants* have lived in the Cumberland region ever since the Cretaceous era when they first appeared in North America, flourishing there throughout the long time required for the first primitive ancestors to evolve into the many kinds of broad-leaved forest trees that we have today. Most of the time the hardwood forest has extended far beyond the limits of the Cumberlands, although now and again its range has been sharply reduced by geological changes; but through repeated cycles of uplift and erosion of the land, always there has remained some area with abundant rainfall, hilly topography, and fertile soil where the many hardwoods might flourish.

Just before the Ice Age began, this forest was widespread over the Appalachians and extended westward into the Mississippi lowland. Except for a reduction in its northern extent, it was little affected by the advent of the ice. Then as the ice withdrew and the climate to the north became more temperate, the oaks and chestnuts and other plants that thrive in similar situations ventured out into the drier, warmer parts of southern New England behind the advancing northerly forest; and there they live today.

As the southern plants moved into a region, they took over first the better habitats. The older residents still persist in bogs or on cliffs, or in other inhospitable places where the newer arrivals do not compete vigorously enough to dislodge them; but as the bogs fill or the cliffs accumulate a soil cover, the out-of-place plants are gradually overwhelmed and disappear.

New England today harbors still another group of plants, also of more southerly connections. These are the holly and southern white cedar, inkberry and rhododendron that are scattered in a spotty fringe along the coast from southeastern Connecticut northward and eastward, some of them reaching as far as Newfoundland.

* Including small and inconspicuous flowers such as catkins, as well as showier "blossoms."

South of New York they live on the broad sandy and swampy coastal plain. But in New England very little of the coastal plain now stands above water. Only along the shore from eastern Connecticut to southern Maine and on the offshore islands are there irregular bits and pieces of flat, sandy country where these plants can grow.

There was a time when New England, too, was bordered by a wide coastal plain. During the height of the Ice Age so much water was piled up on the land in the form of ice that all the oceans of the earth were several hundred feet shallower. As a result, a large area of the continental shelf was exposed in what is now offshore New England. There was an appreciable interval of time after the ice had melted back far enough to expose the coastal plain and before the sea rose with meltwater and flooded in over it again. In that time the coastal plain plants migrated northward over a broad zone of their preferred kind of lowland. Since then the rising sea has cut them back drastically and left only isolated patches of them here and there.

Conditions have continued to change climatically. Evidence from peat deposits in various parts of North America and Europe show that after the last retreat of the ice, climates everywhere became progressively warmer, reaching a maximum between six and four thousand years ago and, with minor fluctuations, have since become somewhat cooler again. The time of greatest warmth was named by Scandinavian geologists the "climatic optimum," and the name has stuck, although a time of greatest warmth and dryness would not be so clearly an optimum in North Africa as it would in Norway.

In the past century average temperatures have been rising, but we do not know enough about the causes of long range fluctuations to tell in advance which changes are minor eddies in the current and which are part of a great and lasting trend.

It may be that we are still in an interglacial or late glacial rather than a truly postglacial period. For most of the earth's history there has been no permanent ice even at the poles. When the ice sheets were at their largest they covered about thirty percent of the earth's land surface. Since the last great thaw set in, the permanently ice-covered area has shrunk greatly; but some ten percent of the world's land is still icebound in the 1950's. It has been pointed out that for a penguin in Antarctica the Ice Age is here and now. It is only man's myopic habit of referring everything to the standard of his own here-and-now that leads us to think that the Pleistocene, the Age of Ice Sheets, has ended.

2

The Scene Changes

TRADITION HAS IT that the first European settlers in America had to chop their way tree by tree into a solid wall of impenetrable forest that reached from high-tide-line of the Atlantic Ocean to the edge of the prairie in Illinois. But our tradition has been so strongly influenced by the heroic labors of the pioneer axman farther west that we have forgotten what primeval New England was like.

Long before the days of colonization, travelers and explorers cruising along the coast found that the forest was interrupted in many places by large tracts of open land. Almost without exception the earliest witnesses commented on the treeless areas they saw anywhere from the Saco River in Maine southward beyond the Hudson and even far up the river valleys into what we know as central New York State. For example, Verrazano, traveling inland from Narragansett Bay in 1524, reported "open plains twenty-five or thirty leagues in extent, entirely free from trees or other hindrances."

In other places the forest was honeycombed with cleared fields where the Indians cultivated their corn and beans. All the Indians of the Northeast were part of the great Algonkian relationship. Those near the coast were to some extent hunters and fishermen, but the staple of their living came from agriculture. These tribes lived in more or less permanent villages of dome-topped wigwams that were built of a wooden framework covered with bark or skin. Around each village lay the cornfields, cleared by the men with fire and stone ax, cultivated by the women with hoes of bone or shell, and fertilized with the well-known fish that was laid in each hill at planting time. Primitive as the operation was, the fields produced large quantities of corn of many kinds, as well as beans, squash, and tobacco. Judging from early accounts, the amount of land under cultivation must have been truly impressive. William Wood, for example, in an early piece of promotional literature published in 1634 and called "New England's Prospect," described one Indian cornfield after another in specifically named places all along the coast.

The same field was cultivated for several years running. When the fertility of the soil eventually declined and the crop yield began to fall off, new land was cleared and the old abandoned, to grow up in time to brush and trees. Trees were also cut lavishly for fuel, and the woods steadily receded from a village as long as the village remained in one place.

Beyond the cultivated lands the forests were repeatedly burned. Many early voyagers spoke of the number of fires they saw all along the eastern coast, and at the same time commented on the open nature of the forest, the large trees "without underwood, and not standing so close but that they may anywhere be rode through." Students of vegetation do not agree as to whether it was fire that kept down the undergrowth, or whether the forest burned readily because it was naturally dry and open for reasons of climate and soil. We do know that the Indians deliberately set fires in order to improve the growth of grass in early spring for the benefit of game animals, and to clear away the underbrush and so make their hunting generally easier. It is also recorded that wet, swampy land, where fires could not easily penetrate, was commonly covered with a dense, thickety growth. So it seems likely that the openness of the forest was at least in part the doing of the Indians.

When the Pilgrim Fathers at last dropped anchor in Plymouth Harbor, the prospect of the bare December woods must have made many a heart sink in secret. One may wonder if they could have survived even as well as they did if, in their weakened, half-starved condition, they had had to make a clearing in the forest before they could put in their first crop the next spring. Fortunately the kind Providence in which they so fervently believed had led them to the site of an abandoned Indian village. Only three or four years earlier the original inhabitants of the place had been virtually wiped out by a plague of smallpox. So it turned out that the Pilgrims found land already cleared for them, as well as a hidden store of grain that carried them through that first grim winter. Then in the spring Squanto, the Indian who was to become their fast friend and helper, appeared from the woods to begin their initiation into the uses of Indian corn.

In the first years there was little occasion for any undertakings beyond securing the necessities of life. But soon more venturesome scouting parties penetrated farther inland. When the first explorers returned from the north, they told with awe of the dense, dark forests of the interior. There there were no sunny fields where women chattered at their work, no open glades where little wild Indians could run and shout, but only an occasional band of hunters passing along narrow woodland trails to break the deep stillness.

Once the first settlements were chipped out of the edge of the wilderness, immigrants came on in increasing numbers. In many cases entire communities came directly from England to start a new life together in the New World. Then as immigration and natural increase brought crowding in the first tight little communities, new towns began to split off from the old and move on to new places. Naturally the most fertile and accessible regions filled up first. By 1700 there were eighty thousand people living in the low-lying areas along the coast and up the great central valley of Connecticut and Massachusetts as far as Northfield. By 1776 Thomas Pownall could write that the land between New Haven and Hartford was "a rich, well cultivated Vale thickly settled & swarming with people. . . . It is as though you were still traveling along one continued town for 70 or 80 miles on end."

As time went on, the flood tide of settlement moved westward and northward, creeping up out of the broad valleys and into the more rigorous, stony hills. Litchfield in the upland of western Connecticut was founded in 1719. Petersham in central Massachusetts was settled between 1733 and 1750, Grafton County, New Hampshire, which includes the western part of the White Mountains, in 1760 to 1772. By 1780 the farthest frontier stood in central Maine and Vermont.

The years between 1830 and 1880 were the heyday of New England agriculture. Probably every American's vision of country life is colored by the family-size farm of those days as it was in the more settled parts of New England. The farm was as nearly as possible a self-contained economy, with useful work for everyone, however large the family might grow. The littlest ones could hold the skein of yarn to be wound for grandmother's knitting, or set out milk for the cats, who in turn earned their keep as mousers. A boy was needed for man's work as soon as his muscles were strong enough. And a woman's life was filled with the routine but tangibly constructive work of sewing and cooking and the eternal details of keeping a family clean and comfortable.

Pleasures were simple and home-made. If the household needed outside distractions, there were plenty of neighbors within gathering distance for barn-raisings or quilting parties. Odell Shepard, writing about Connecticut, said, "In early days our people could see the lights in one another's windows and could communicate by shouting from farm to farm. An old man has told me that in his youth it was possible to arrange a barn dance among a dozen of his neighbors without any man's stirring from his front door. . . . Another old man has told me of the time, far back, when he, sitting under his elm tree on a Sunday afternoon, could see seven or eight friends of his sitting in front of their several houses and under their own elms." * For the most sobersided there was always

* In *Connecticut, Past and Present.* Published by Alfred A. Knopf, Inc.

Sunday church meeting and the excitement of occasional camp-meeting revivals.

We have an outsider's picture of the times from Harriet Martineau, an English "authoress" who traveled widely in America in 1834-36 and wrote a charming and lively account of what she saw and what she thought about it. One of the houses where she visited stood on a hilltop in Stockbridge, a handsome old town in the Berkshires of Massachusetts. From her host's doorway she could see everything that went on in the busy village, since, apart from the forest-covered mountains, the entire landscape consisted of lush green cultivated fields and pastures, broken only by the road and the Housatonic River. She also visited Gloucester, and thought that "the place has the air of prosperity that gladdens the eye wherever it turns, in New England;" and she observed (as her contemporary Karl Marx, sitting for years in the reading room of the British Museum, never realized) that "as a mere matter of convenience, it is shorter and easier to obtain property by enterprise and labour in the United States, than by pulling down the wealthy."

Although the first aim of a farm in those days was to provide the family's material needs, a cash crop of some sort was needed on even the meagerest place to provide for a few necessities such as iron tools and salt; and many amenities to make life more pleasant were available for the buying. Most of the scant plowable land was needed for subsistence crops, but there was plenty of steep and stony land that made satisfactory grazing for cattle as soon as it was more or less cleared of trees. Over the years, consequently, more and more cattle were raised for a market crop, and from the beginning an increasing proportion of farmland was devoted to hay and pasture. In the 1840's, for example, about eighty-five percent of the town of Petersham was cleared of forest, and by far the most of this was used as pasture.

Even in the midst of this happy summer season, and before the frontier tide had reached the northern limits of the country, the ebb had begun almost unnoticed in the earlier settled regions. It was the little towns in the hills that slipped first. It took only a generation or two of life on a high, remote farm, where summers are short and winters long and bitter, and where the sheep need sharp noses to graze between the rocks, to send a farmer off looking for a more promising place. There is the case of the Connecticut hill town of Hartland, scarcely twenty miles from Hartford. The population of Hartland township has shown a decrease in every federal census since the first one in 1790. The condition soon spread; and as early as 1820 many towns in rural parts of southern Vermont and New Hampshire had begun to lose population.

For a long time the trend of migration from the ebbing districts was northward into the still empty parts of upper New England. Gradually, however, an even stronger tide to the west set in as the front line of advancing settlement poured out onto the fertile, level lands beyond the Alleghanies. The flood gates to the Midwest really swung wide with the opening of the Erie Canal in 1825 and the beginning of steam navigation on the Great Lakes in the 1830's.

Contrary to what many people believe, there is nothing inherently wrong with the fertility of New England's soil—what there is of it. The highest yield of corn per acre produced in this country until recently was produced in Connecticut. The difficulty lies in what the glacier did to the topsoil, and the key to the matter is the phrase "what there is of it." All the soil that once mantled the landscape was scraped away by the ice. Some of it was carried off and dumped in the ocean, and the rest was thoroughly mixed with rocks of all shapes and sizes and of great abundance before it was set down again. Fortunately there were rivers and lakes that washed some of the soil out from among the rocks and assembled it into usable masses here and there. But relatively few of these patches are large enough to do more than provide turning space for a small horse-drawn hayrake.

It was the competition from cheap land, level and clear enough to allow the use of large farm machinery, that put the pinch on New England agriculture. When canals and then railroads came along and provided low-cost transportation for bulk freight from the West, the bottom fell out of the old farm economy. As a result, hordes of Yankees gave up and went off to populate the new lands; and it is not always as easy as one might think to tell an old-stock Ohioan or Iowan from an old-stock Vermonter.

Defection from the hills received a further push from the expansion of water-powered industries, and this in turn was enormously stimulated by the Civil War. People who did not go west moved down into the mushrooming factory towns nearer home. While the farmer's daughters went to work in the mills, his sons went off to fight in the war. Perhaps it was the general restlessness of soldiers going back to civilian life. Perhaps it was just seeing other parts of the country and other ways of living that offered greater rewards for toil. In any case, large numbers of young men never returned to the old hill farms.

By the 1870's farms were being abandoned wholesale, even in the recently opened parts of northern New Hampshire and Maine. Deserted farmhouses became increasingly conspicuous in the landscape, and soon it was apparent to even the least observant that a great change was taking place in rural New England. The general public grew highly excited, and a loud cry of alarm went up over the decline of a way of life that had become centrally embedded in our national tradition. Files of popular magazines of the 1880's and 1890's show the state of public opinion. Every volume for those years has articles written from all points of view, impassioned, reasoned, or merely sentimental, setting forth proposals for keeping people on the farms in order to preserve our Great Heritage of plain living and high thinking, and of course in an idyllic rural setting.

The farmers who were trying to squeeze a decent living from the rocky hills took a different view of the matter. When a family decided to leave, there were few takers for the farm. Many simply moved out and, after a last lingering look at the old home, shut the door and went away, leaving the place to the forces of nature.

With no one on hand to repair a leaky roof or replace the first broken window, it took only a few years for an abandoned house to fall into decay. No one was there to grieve over the more complete havoc wrought by fire or violent windstorm. With the garden unweeded, the paths untrod, even the pastures ungrazed, the land that had been so laboriously cleared soon grew up to brush, its very existence as a homesite all but forgotten. In less than a generation there might be nothing left but a cellar hole far in the woods on a road no longer kept up by the town. A man from southwestern New Hampshire once said that when he was a child in 1865 he knew of nine old cellar holes within a mile of his country school. In the same area in 1887 he counted twenty-three of them.

The course of farm abandonment in New England can be read in the changing proportions of cleared and wooded land. By 1860, all but twenty-seven percent of the entire area of Connecticut was open land. Those were the peak years. By 1910 the woods had expanded to cover forty-five percent, and by 1955 sixty-three percent of the state. Farther north the cleared area never reached such an extent, and the maximum came later. Only twenty-five percent of Maine has ever been open country, and that was in 1880. At the present time about three-quarters of all New England is covered with woodland or forest, and the remaining quarter includes all the cities and their sprawling suburbs.

Other hilly parts of the Northeast—New York, Pennsylvania, northern New Jersey—have a somewhat similar history of agricultural development and retrenchment. But in none of these regions was either the flood tide or the ebb so strong as it was in New England. All those states have enough flat or gently rolling land that is good for farming even under modern conditions so that by the time the good land was all taken up and settlement began to lap against the hillsides, the main impetus of frontier expansion had already moved on to the Midwest.

When the woods begin to reclaim their old dominion, the course of events follows a fairly well-defined pattern. Of the many kinds of plants that are always ready and eager to take over any undefended living space, the first to infiltrate an untended field are those that "get there firstest with the mostest." For an invading plant, this means the production of an abundance of seeds that are readily and widely dispersed and that can grow rapidly into vigorous seedlings. For this purpose, seeds may be light and easily carried on the wind, or they may be borne in soft fruits or berries so that they will be eaten and scattered by birds. Then they must grow into young plants that will thrive among the grass and weeds of an old field and be able to withstand the rigors of full exposure to hot sun and drying wind. Each region of the country has its characteristic plants that meet these requirements.

In the southern and southeastern parts of New England, encroachment by the woods begins with the appearance of tiny red cedars scattered among the dry, skimpy grass and goldenrod of a neglected pasture. Year by year the cedars grow taller, but they keep their slender forms and rarely become thick enough to make a dense forest. At a certain stage they may look like an army of dark-clad soldiers deployed in loose formation among the boulders of a broad hillside. With the cedars come scraggly little gray birches and sometimes rank-growing black cherries, beloved of the tent caterpillar. A common sight is an old veteran of a cedar that has served so long as a bird perch that it has become surrounded by a thicket of young cherries.

In the extreme north and northeast and at higher elevations elsewhere, the edging-in of red spruce, with a spatter of balsam fir, betrays the deterioration of pasture land.

Throughout the great central region, however, it is white pine that first takes over the old fields. Even today, pines are so characteristic of the region that for a vignette symbolic of New England, one might use a rough, bouldery hillside covered with short grass and clumps of fern, a bit of stone wall, and a single ancient pine, gnarled of trunk, horizontal of bough, and soft and delicate of foliage.

At the height of agricultural development many old pines survived in woodlots and along roadside walls and in pastures, where they were left to provide shade and shelter for the animals. In a good seed year an old pine produces seeds by the million; and being light, and each equipped with its little sail, these easily blow out across the fields. The open, sunny grassland of an old pasture suits an infant pine very well, and in a few years an unused field becomes thickly stocked with thrifty young trees. By the time these are head high to a grown man, their branches may meet in a completely closed canopy that darkens the ground below and smothers the old pasture grass and weeds. In the deep shade and in the dense, springy mat of fallen needles that soon forms under a pinewood, few other plants can get a foothold; and the pine grows up in a virtually pure stand, with only scattered wisps of rather half-hearted underbrush.

As the trees grow older and larger, and the thick canopy of foliage rises, gradually some light and space begin to appear below it. Heavily shaded lower branches die off, leaving an open chamber, brown carpeted and punctuated by tall dark trunks with leafless side branches. Occasional breaks appear in the ceiling where a tree succumbs to windstorm or blight. Such islands of light are quickly floored over by a brushy thicket. Elsewhere the woods are dark and quiet. No sun-loving pine seedling can survive in the dimness; but gradually shade-tolerant Christmas ferns and partridgeberry and ladyslippers begin to infiltrate, along with scattered seedlings of red oak, cherry, and hard maple. The invasion is sparse and slow at first, but persistent.

By the time the old-field pines are some fifty to eighty years old, the leafy underbrush is becoming conspicuous, and it is clear to a thoughtful observer that the next generation in these woods is going to be different. At about this time, too, the growth of the pines begins to taper off, and after another twenty years or so their growth becomes extremely slow. From then on almost anything that can happen in the forest causes death and destruction of the pines, whether windfall or fire, disease epidemic or insect plague. Every pine that goes leaves space for some shade-tolerant sapling that has been inching along in the dimness, awaiting its day in the sun. Before many years the remaining pines are being crowded out by the burgeoning hardwoods, and the days of the pine forest are numbered.

When they are full grown and just before the younger hardwoods begin to crowd them, the pines are ripe for lumbering. Although most of their wood is knotty and by no means up to the quality of slowly grown virgin timber, still it is useful for boxes and crates and matchsticks; and when there is a market nearby, it can provide a tidy return for both land owner and lumberman.

It was between 1890 and and 1925 that the great harvest of old-field pine was reaped in New England. In those years lumbering was done by clear-cutting. This is essentially a mowing operation in which everything is cut off close to the ground. Large trees are lopped of their branches and cut into sawlogs. The largest branches and smaller trees are cut up for posts or cordwood. Scraps may simply be left to decay where they fall, meanwhile providing shelter for small creatures of all kinds, including some of the most troublesome insect pests, and drying into the most inflammable kind of tinder. Or the "slash" may be gathered together into heaps or windrows and burned on some still, damp day as a precaution against disaster.

In either case a cut-over forest presents a desolate picture. Scattered among the raw stumps are ferns and a few little flowering plants, now suddenly deprived of their overhead shelter. Small tree seedlings and saplings that are flexible enough to bend rather than break under a skidding log will remain, if somewhat the worse for wear. The soft mat of fallen needles is churned up where the logging horses struggled for a foothold with their heavy load, or in modern times where grinding truck and tractor wheels dug into the earth. Mineral soil may be exposed by wear of the traffic passing back and forth in the trackways, inviting erosion and the start of gullying in the next heavy rain.

The climate in which the forest survivors live is abruptly and drastically changed. Now they are exposed to the full blast of hot, drying sun and wind. There is no protecting roof to break the force of pounding rain or to hold an insulating blanket of warm air against the chill of a clear, frosty night. The new conditions are too much for some of the forest plants, and many of them languish and die.

For others, though, the going of the pines brings a new lease of life. Youngsters of oak, maple, and beech go into a rapid spurt of growth and soon make a thick woodland. In the old days chestnut was an important part of the new growth. Any of these that were large enough to be cut in the lumbering operation would sprout vigorously from the stump, growing all the more rankly for the large underground root system supplying the new shoots. White pines, on the other hand, like most other cone-bearers, do not sprout, and a tree cut down is a tree gone forever. Moreover, any pine seedling that starts among fast-growing stump sprouts is shaded to death in its infancy. So wherever hardwoods have a firm foothold they effectively shut out any new pines, and the next generation of forest bears little resemblance to the old.

Any parts of the old forest that were not yet infiltrated by hardwoods at the time of lumbering soon become restocked with seedlings of one kind or another. If a good pine seed year follows directly after the cutting operation, and if there are enough old seed trees left in the neighborhood, the new stock may consist at least partly of pine. Chances are that there will also be plenty of light-seeded or bird-distributed birch, poplar and cherry. Even when outnumbered by pines, the hardwood seedlings grow so much faster in their first few years of life that they offer strenuous competition to the new generation of pine. Here things are quite different from the dense sod of an old pasture, where the pine is much less inhibited than the hardwoods by a close company of grass and weeds. As a result, the new generation of woodland is promptly dominated by broad-leaved trees, and the old-field pine goes the way of all transients.

In the thirty to seventy years since the great pine harvest, landowners have taken from the new growth whatever they could find that was useful or salable. Most often this has meant repeated clear-cutting for cordwood. The little selective logging that has gone on has removed the better trees of kinds that could be sold for sawlogs, leaving the trash to develop for the future. Most of our hardwoods will sprout from the stump; but though sprouts from a very small, cut-off tree may grow into quite respectable timber, those from a large old stump never form usable logs. Moreover, the most vigorous sprouters are not the best timber trees. As a result, the "forest" cover of much of central and southern New England is now a sorry mixture of the most persistent weed trees and low-grade stump sprouts.

Much is known that could vastly increase the value of the forest output if it were ever put into wide practice. As it is now, low-lying, swampy land is most often covered with almost pure red maple that is good for nothing but an indifferent sort of cordwood, so much so that "swale" means "red maple" to many local farmers. Yet experiments have shown that with a minimum of weeding, thinning, and improvement cutting, such a swale can be converted in twenty-five years to a stand of straight, well-formed trees of good species such as ash and yellow birch that can be used for lumber.

On heavy, fertile soil that naturally supports oak, birch, beech, and sugar maple, a simple silvicultural program of weeding, thinning and pruning will produce a timber crop far superior to the kind most often found in such places today.

In the many places where the soil is light and sandy or gravelly, hardwoods do not grow so rampantly, and the less demanding pines can keep the upper hand. New England has countless flats and benchlands made of the coarse, quick-drying materials that many glacial lakes and streams deposited. Here forests of pine no doubt stood for thousands of years and will return with very little encouragement. These will be our pinewoods of the future, when old-field pine has become a legendary thing of the past.

When all the trees are cut at one time from such poor, sandy places, leaving a mat of old needles exposed to wind and sunlight, the soil may quickly become covered over with a stubborn, durable ground cover that strongly resists the invasion of trees of any kind, even the drought-tolerant pines. Any New Englander, native or transient, knows such dry, scraggly wastes of low-growing blueberry, dotted with patches of moss and reindeer lichen, with sweetfern adding its scent to the fragrance of the nearby pines in the hot summer sun.

The blueberry that so delights the summer picnicker is a tough and resilient pest to anyone who wants to put his land to some kind of more productive use. For an example, there are great tracts of dense blueberry thicket in the Pachaug State Forest in eastern Connecticut; and it is a problem just how this scrubby wasteland can be converted back into woods. Foresters who were sent in to study the situation recommended planting pines, commenting that, though the pines will start slowly there and won't improve the already poor soil, they will in time yield a valuable crop; and if the pines can eventually crowd out the blueberries, "the net result may be beneficial."

Of course not all the land once cleared for farming has been abandoned to grow up to forest or blueberry barrens. New England agriculture has not dried up but rather it has taken other channels. Recent decades have brought a shift to highly specialized farming; and there are countless part-time farmers who contribute a substantial share of the agricultural output of the region.

Under the term "rural non-farm population," the federal census lists other thousands of happy people who are enjoying country scenes with the comfort of city incomes, and whose "farm," as they will call it, may consist of little more than a few decrepit but delectable apple trees and a few acres of grassy field that are mowed more to keep the brush down than for the sake of the few wisps of hay they may yield. The houses that time passed by and the villages where nothing much has happened for a hundred years may bespeak a functionally decadent landscape; but they please our eye and give us a sense of having roots. And today's non-farmers can look with pleasure on the rocky, wooded hills that so discouraged their great-grandfathers, the spiritual if not the literal ancestors of us all.

3

The Famous New England Weather

MENTION OF New England weather brings to mind sayings like, "If you don't like the weather, wait a minute," and "We have two seasons, July and winter," or its variant, "Nine months of winter and three of rough sledding." Then there are those chains of buildings—house, summer kitchen, woodshed, barn—all interconnected so that the farmer need not be isolated from his dependent beasts whenever the household is snowed in. The outlander may well wonder just what the New England climate is really like.

The answer depends on what part of the region he is wondering about. Although New England is not large compared to other parts of the country, it suffers a fairly wide range of climate. The southeast corner, tempered by ocean winds, resembles places much farther south. The top of Mount Washington is about like Labrador. Still, there are a few characteristics that distinguish New England weatherwise from other parts of North America.

For one thing, there is no dry season. Throughout the year precipitation averages between three and four inches a month, although there are occasional months with five times that amount and others with none at all. During the spring and summer the rainfall is about the same in New England as it is in the rest of the northeastern states from Virginia to Iowa. But whereas autumn and winter are substantially drier in other places, in New England the moisture keeps coming; and on the average it rains, snows or sleets on about one day out of three the year around.

In spite of the abundant precipitation, there are about as many really clear days here as there are in Iowa or in most of the Deep South, including Florida. In Connecticut, at least one day in three is bright and sunny, although the expectation in northern Vermont and New Hampshire is substantially less.

On days that are warm and sunny everywhere else, fog may lie chill and damp along the seacoast, especially from Cape Cod northward. This happens when warm, moist air moving up from the south or in from over the Gulf Stream hovers over the cold ocean and becomes cooled enough so that its water vapor condenses into a cloud of microscopic droplets. Sometimes one can see the gray line of fog approaching from far out at sea and watch it drift in over the land until boats and beaches, rocks and trees are shrouded in a thick, clammy whiteness, and the clang of bellbuoy and mooing of foghorn give their own flavor to the day. Yet often on such a day one need walk only a few hundred yards inland or around behind a sheltering headland to emerge into a world of warm sunshine and blue skies.

Thick fog may form in the inland valleys, too, especially in May and September when noons are warm and nights are chilly. This kind of fog materializes out of thin, clear air during the evening and drips from trees and bushes all through the quiet night. Early risers the next morning know that if the fog lifts from below it will come down again later as rain; but if it lies close to the ground, it will "burn off" from the top to make a fine day. On such a morning, hilltops emerge first into the sunlight, and the last foggy remnant may lie late in the valley bottom like a fat white roll of insubstantial cotton.

The New England climate offers nothing so violent as a Texas norther, a Kansas heat wave, or a Dakota blizzard; but the weather fluctuates much more often within its lesser range of possibilities than it does in most other parts of this country. A glance at a map of storm tracks across North America shows part of the reason for this changeableness. Our storms follow two main paths. One of these comes in from west to east across the Great Lakes; the other comes up from the south along the Atlantic coast; and both go out to sea over New England and the Saint Lawrence Valley. Moreover, New England lies just athwart the zone where cold dry air from the arctic regions over Canada encounters warm, moist air moving up from over tropical seas. These two very different kinds of air mass are constantly jockeying back and forth and, when they collide, set up great turbulence.

The resulting weather offers many variations on a basically constant pattern. One day the wind blows cool and dry out of the northwest, and the air is so transparently crystalline that every blade of grass stands out individually on the farthest hill. The next day may look and feel the same, but the weather-minded outdoorsman, noting that the wind is in the northeast and the sky brushed with the plumy mares' tails of cirrus clouds, darkly labels it a "weather-breeder." Sure enough, the barometer falls, and a day or two later brings a deluge of blowing rain. And

Rain from the east
Rains three days at least

—though not always so in the summer. Or perhaps the wind shifts from northwest to southwest and the forecast is "fair and warmer." This happy state may last day after day, but eventually the fair and warm turns to hot and muggy, and the rain clouds that shut out the oppressive sun come as a welcome relief. In this case, too, the mass of rain-spilling air lies for several days, until eventually the wind veers once more to the northwest. Even while it is still raining, the weather-sensitive person can feel his spirits and his energy rising with the barometer; and keeping a watchful eye on the western horizon, he may be rewarded with the sight of a sharp cloud edge pulling off toward the east like an opening curtain to let the sun in on a new-washed world, rounding one more cycle in what the weather bureau describes as "a more or less regular succession of bi-weekly storms of snow or rain, with intervening two- or three-day periods of fair weather."

Summers on the whole are moderate in New England. July temperatures are like those in northern Michigan or Minnesota. The hottest summer days usually reach ninety or ninety-five degrees, comparable to northern Wisconsin or Minnesota or the lower Colorado Mountains. It is hottest oftenest in central Massachusetts. Restaurants and movie theaters in any distinctly valley region, even the Champlain Valley of northern Vermont, are air conditioned. Really hot nights are rare, however. After-dark swimming is practically unheard of, except by the young and dauntless, and one may regret leaving home for the evening without taking a wrap of some description.

But what is an old-fashioned New England winter like? The relatively few people who live on the northern fringe from Vermont across into Maine can expect it to be both long and cold. In January the temperature averages lower than ten degrees, and on an occasional night it drops to thirty below zero. Snow often lies on the ground continuously from Thanksgiving to mid-April, and over eight feet of it usually falls during this time. Once at Vanceboro, Maine, eight feet of snow fell in the space of four days.

Even allowing for the effects of a January thaw and one or two other mild intervals, snow is a factor to be reckoned with all winter long. It has to be shoveled from walks and doorways and even from rooftops. In exceptionally snowy years many an old barn slumps over on its side under the weight of it. Roads soon run through blue-white canyons whose walls grow higher and higher as the plows throw each new snowfall up on top of the last. Nowadays people are rarely snowed in, even for a short time, since highway crews are out with their powerful plows almost as the first flakes fall.

Then some dim, gray morning after months of constant snow and ice, people in the north country waken with a vague sense of pleasant anticipation, wondering why the mood so poorly matches the sodden day. Finally it rises to the level of consciousness—the almost forgotten sound of rainwater trickling in the eaves troughs, raising thoughts and hopes of spring.

The spring thaw in such northern country is a formidable thing indeed; and when bare soil at last shows through the slushy remnants, it marks the beginning of Mud Season. During the long winter, as snow piles up on top of the earth, the cold penetrates several feet down into it and freezes the moisture that is abundantly present there in late autumn. With the onset of spring it takes several weeks for the frozen earth to thaw from the top down, so that snowmelt, as well as newly falling rain, can drain away from below. Meanwhile the thawing surface is an undrained, muddy morass. The clean, dry snow of winter is much easier to cope with than this; and those who can go off on winter vacations usually stay home until just before spring and then go away until Mud Season is over.

Near the coast in the southeast, winter is quite a different matter. There the long, mild autumn lasts well into November. One rarely need think about heavy winter coats until after Thanksgiving Day, and the first snow may not come until almost Christmas. Sub-zero weather is almost unheard of. Most of the time it will freeze by night and thaw by day, and most snowfalls either begin or end as rain. It is, alas, a land of sleet and slush, of the alternate freezing and thawing that tears up pavements and heaves garden plants out of the soil. But it ends soon enough. By mid-March song sparrows are proclaiming the coming of spring, and in another two or three weeks the grass turns green and buds begin to swell.

In the central region where most people live, winter is pleasantly crisp and not too everlastingly long. January there is about as cold as it is in Ohio or Nebraska. There is enough snow to keep the ground cleanly covered, and it stays cold enough to keep the snow frozen most of the time. This is the land of Whittier's "Snow-bound."

Since 1938 the ordinary New Englander turns pale at the mention of the word "hurricane." Actually, though we knew less about them in days when communications were less highly developed, hurricanes have been blowing into New England for centuries. Exceptionally large and ferocious tropical storms struck here in 1635 and again in 1815. From the records, it appears that five to ten hurricane-type storms reach New England in a century. Fortunately most of them are much less devastating than the terror of 1938.

Floridians are likely to be scornful of the Yankee's emotional reaction to a hurricane. But consider the situation of the coast of Connecticut and Rhode Island. This east-west shoreline lies exposed broadside to any storm coming up from the tropics. If a hurricane travels northward through a region already overlain with warm, moist air, and if it remains over the open ocean all the way from the West Indies to New England, it can continue to pick up force all the way to its northern landfall and hit there with undiminished fury.

The New England coast is very ragged and irregular, with many inlets, embayments, and estuaries reaching in between the hills. The low-lying shores are thickly occupied by man and the works he has built for his business, industry, and pleasure. When an already high tide is made higher by the sharp drop in barometric pressure that accompanies a great storm, and when this tremendous tide is driven inland by a following wind of hurricane force, the flood surges far up into all the coastal indentations and up over all the adjacent low terrain. Wharves and warehouses, factories and homes suffer even more from the pounding of storm waves and flooding by high tides than from the violence of the wind.

In this heavily populated region, many sandspits and beaches along the shore are thickly built up with summer and even year-round homes. The long, narrow sandspits formed by complex currents off the irregular shore offer little resistance to a hurricane tide; and in 1938 many such, along with their entire seaside communities, simply vanished into the sea. So it is small wonder that Yankees stand in awe of a hurricane.

Still, New England's weather rarely indulges in such dramatic behavior. Normally it is a thing of constantly changing details, each to be savored while it lasts, whether a snapping, starlit winter night, a spring day of petulant showers swept with flashes of sunshine, or a long, golden summer evening such as only a spare, northern land can produce.

4

The Fields and Their Lilies

STRANGE AS it may seem at first thought, the climate of New England is actually better for growing grass than is the climate of the natural grasslands. On the prairies and plains grass can grow only during the spring and early summer months, when most of the year's precipitation falls in those regions, and must suspend operations through periods of heat or cold or drought. This the grasses are quite able to do; but trees cannot survive the prolonged dry seasons there and so the more tolerant grasses have the land to themselves. In the relatively cool, moist weather of New England, on the other hand, grass can continue to grow through the summer and late into the autumn. But our abundant year-round moisture also allows the growth of forest; and where trees can grow, grass cannot by itself hold out against their competition

In the primeval wilderness of New England naturally open treeless places were few indeed. To be sure, the Indians cleared and cultivated substantial tracts of land; but as soon as a field was abandoned, it rapidly grew up to brush and was soon recaptured by the trees. So scarce were the natural grassy places that when the Founding Fathers arrived from England one of their most immediate and pressing needs was to find pasture and hay for their livestock. In fact, grass was so important that land grants, drawn up perhaps far from the scene in a very different kind of world, often specified just how much meadow the allotted acreage should include.

Immediately along the coast the only available grass grew in the tidal marshes. The woods provided a certain amount of grazing in season, but hay for the long winter at first came only from the marshes. Even after clearing had let in the light so that upland fields could grow a cover of grass and the other little plants that cattle eat and grow fat on, salt hay was a staple for winter feeding. It was cut in late summer and stored in hay barns that were built on some rise of land conveniently near the mowing but standing safely above the highest rise of winter storm tides. Most of the buildings are long since gone; but here and there one leaves its echo in the suggestion of a trackway leading toward the site, or in some rocky hummock that keeps its old name of Barn Island.

The many stone walls that to the city-bred eye reach out to seaward over the marsh in such a meaningless way really testify to the importance of clearly defined property lines on land as valuable as good hay marsh was in the early days. Even now salt hay is worth harvesting, not so much for fodder as for packing and mulching and other uses where its firm springiness makes it superior to other materials. There are marshes that have been mowed for hay almost every year for three centuries, right down to the present time.

The best natural grasslands, however, were river meadows. Pioneer settlement followed wherever they led. Accessible places like the great meadows near Hartford and Springfield and along the lower Merrimac River filled up with a dense and thriving population long before 1700. Other little pockets of rich, grassy bottomland, even intervales tucked away far from the main body of settlement, had their small hamlets or isolated farms when the nearby upland was still a silent, empty wilderness.

In these natural grasslands it is primarily wetness that keeps out the trees. Tidal marshes are flooded twice every day, and with salt water at that. None of the trees that grow in our climate can live under such conditions. In other low-lying places flooding is seasonal. Some of our trees can tolerate shallow standing water in winter and early spring, when their inactive roots have less need of abundant air. These are the ones that spill down from the hills to make swamp forests on poorly drained flats and hollows. But where standing water is never far below ground level, and where swirling floods wash at the roots every year, the land belongs to the shallow-rooted and tenacious grasses.

In other places the soil is too dry for our trees, and drought-tolerant grasses and other adaptable small plants have things to themselves. Even in this well-watered land coarse, sandy terraces standing above the general underground water table drain quickly from below after every rain or thaw, and the top layers of the soil are almost constantly dry. Once the native plant cover is broken on such a sandbank, the wind can tear deep into the loose, dry mass to make a rapidly growing "blow-out." One enterprising Yankee saw the possibilities inherent in the situation; and when, in spite of years of effort to arrest it, such a blow-out on his farm only grew larger and larger, he gave up the struggle and began to exhibit the sandy waste to tourists as "The Desert of Maine."

Other sandy barrens have been less successfully exploited on their merits. There is a big, flat expanse along the Quinnipiac River north of New Haven that remained largely an empty wasteland until recent years, when the growing cities and towns nearby have sent their tide of new houses washing out over the sand plains. Even among lawns and shopping centers, the traveler can recognize the area by its flatness, its sandiness, with not a rock in sight, and the predominance of oaks.

Parts of this sand plain that have not been taken over for roads and buildings look much as they did more than a hundred years ago. Where it has not been disturbed the soil is thinly covered by a sparse growth of bunch grass, with patches of moss and lichen in between. Dotted over the plain are clumps of oaks and a few pitch pines. Such a dry, barren scene is not what one would expect in this climate, especially so close to some of the region's most fertile farms. It has excited the curiosity of scientists, and a detailed study of the area was made some years ago before the suburbs had encroached.

There are three kinds of evidence that explain what a place used to be like and why it is what it is now. These are its present vegetation, its recorded history, and the structure of its soil. All three veins were worked for the North Haven sand plains.

To begin at the beginning of the story, the soil, like so much of New England, is the work of the glacier. Digging down into the earth, one finds that the sand is not homogeneous, but consists for many feet down of horizontal layers of varying textures, though all basically coarse and sandy. This is the sign of an old lake bottom, built up over the ages by sediments washed in from the country around. Like so many in postglacial New England, the lake was long and narrow, caught between the surrounding hillsides and a remnant tongue of ice lying in the valley bottom. Currents flowing through the lake must have been fast enough at times to carry away all the fine mud and silt, leaving behind only the coarser, heavier sand.

When at last the ice was entirely gone and plants could return to the valley, the old lake floor was left as a wide terrace standing high and dry above the little river. There was plenty of rain, but it trickled away so fast through the loose sand that the earth became dry again soon after every storm or shower. Only highly drought-resistant plants could grow in such a place; and for ages the sand plains were covered entirely with grass.

Over the years the grass left its mark on the soil. Even today in spots that have not been disturbed, the top eight inches or so has the same dark rich brown color that one finds in any permanent grassland, whether prairie, steppe, or meadow. It is the darkness of humus, imparted by grass roots growing and decaying in the same place for centuries on end. The dark layer here is not very deep, but it is quite characteristic of a grassland kind of soil, and it gives no indication that any trees or other large plants grew there before the land was cultivated. For two feet below the dark layer the soil is stained a lighter reddish-yellow color by materials washed through from above by filtering rainwater. Still deeper the coarse sand is a pale tan color, very little changed from the time when it was laid down in the glacial lake thousands of years ago.

Old town records of North Haven show that the area was allotted for settlement beginning about 1710. Its grass cover was much inferior to damp meadow for pasture, and it appears to have been used instead for plowland, fenced off into relatively small rectangular fields that were probably planted to grain. There were no rocks at hand for wall-building; and early travelers through the region commented on the general use there of "Virginia fence," the zigzag rail fence that was common farther south and west, where surface stones are less overwhelmingly abundant than they are in most of New England. In certain places one can still find the groups of round stones that were used to support the angle posts, regularly spaced according to the length of the split rails.

Once the original grass cover of the earth was broken by plowing, other kinds of plants could get a start, and the field edges, somewhat protected by their fences, soon grew up to brush. Many trees started from seeds that were blown in across the fields or were dropped by the many birds that live in such sheltering tangles. Farmers ordinarily left many of the existing fencerow trees standing and often planted others of kinds that were especially useful for posts or other purposes around the farm, making a sort of running woodlot. Since fencerows are often very little disturbed for long periods of time, they are likely to contain good samples of the local wild plants. Especially in the more intensely cultivated parts of this country farther west, the "fenceline relict" is often an important clue to the original vegetation of a place.

Such light soil as these sand plains, poor to begin with, was soon greatly impoverished, and by 1800 cultivation was largely abandoned. For a century and a half after that the worthless land was left to develop as it would, at the mercy of natural forces.

Even today some of the old fencerows can still be traced by the straight lines of trees that grew in the angles where zig met zag, though the rails have long since fallen into humus. Between the fencerows much of the area is still open and treeless. In some places wind erosion has blown away patches of the fine, dark topsoil that had formed under the long influence of grassroots, and the heavier sand particles have piled up into a few low-grade dunes. This probably happened when the fields were first abandoned and the soil lay bare. The exposed sandy patches are so inhospitable for plants that only scattered clumps of bunch grass, along with a few mosses and lichens, have succeeded in getting a toehold, leaving much bare soil between.

Less desperately barren spots have reverted to a poor sort of dry grassland. This is highly susceptible to fire for much of the year while the plants are dormant and only their dead leaves stand above the soil. Burning the tops does not kill the underground parts of grass plants; but fires destroy any accumulated humus, further impoverishing the soil, and kill less fire-resistant kinds of plants. Consequently, repeated burning in time produces a stand of almost pure grass. Apparently fires were frequent enough here in pre-settlement times to maintain the grass cover intact for centuries.

Where the grass has been free of fire for several years, tree seedlings begin to appear. The clumps of black oak that are scattered over some of the fields apparently owe their origin to the hoarding instincts of squirrels. The acorn supply is large, and the little beasts hide far more than they ever come back to claim; so they act as highly effective planting machines. Careful digging has turned up caches of acorns in all stages of development, from newly gathered nuts through sprouted seedlings to sizable trees. When the young trees are up above the grasstops, they make patches of shelter from the scorching sun where other plants can get a start, and soon a clump of shrubby vegetation grows beneath and a little to the north of each group of trees where its shadow breaks the fiercest midday sunshine. In these sheltered spots the soil remains moist for a long enough time after each rain to allow other seeds to germinate, and slowly the brushy island is enlarged.

Other kinds of trees come into the grass from fencerows and the surrounding woods, but few of them can stand the harsh conditions.

Pitch pine seeds blowing out into the open in large numbers at just the right season may be able to start a little grove. But in this locality pine tends to be supplanted by oak in the course of time, and the general scene remains one of oak and grass.

There are similar sandy barrens on valley flats and terraces scattered over the face of New England. All of them stem from some old postglacial lake bed. They are of no use for cultivation and not much better for pasture; and whether they have grown up to grass or to oak and pitch pine, they have a strange, discordant look in such a predominantly lush, green countryside.

Salt marsh, river meadow and sand plain—these were the primeval open spaces of New England. The forest was threaded with brooks that occasionally broadened out enough to let in the moderate light that violets and bluets need. Here and there in the mountains were stony screes so steep and unstable that no tree could grow large enough to shade its neighbors before it went skidding down the slope. Many large bedrock ledges and hummocks were scraped so clear of soil by the glacier that only small plants could find root space in the crevices and nothing ever grew big enough to cast much shade. But the natural openings were relatively small and scattered. Most of the land was dominated by trees, and the prevailing color was green.

The modern landscape has far more bright flowers than there were in the primitive wilderness. Few of our native plants will flourish in open fields and along sunny roadsides where the soil is neither very wet nor very dry, and the flowers that grow there are nearly all immigrants. Botanizing in these places with the help of a manual, one comes time and again on the phrase, "naturalized from Europe." A short list exhausts the native New Englanders that are both common and conspicuous among the field flowers: milkweed, cranesbill, robin's plantain, steeplebush, asters, goldenrod. Even the black-eyed Susan came from natural grasslands farther west. The rest—daisies, most buttercups, bouncing Bet, clover, Queen Anne's lace, hawkweed, chickory, the common dandelion, and all the others have followed in the wake of European man.

Whether you look on all these as wild flowers or as weeds depends, of course, on whether your interest is aesthetic or utilitarian. A cow among the daisies is a picturesque sight, but she is not likely to be eating a balanced and nutritious diet; and a farmer casts a jaundiced eye on a field lacy with mustard and wild carrot. Perhaps the best definition of a weed is "a plant where it is not wanted," although the term usually implies also the ability to live exuberantly on scant resources.

However you define them, some of our weeds doubtless came over on the Mayflower. The first horses and cows to land in the New World, along with remnants of the feedstuffs needed to sustain them on the sea voyage, must have brought in domestic weed seeds from the old country. As early as 1672 John Josselyn recorded some forty kinds of weed "sprung up since the English planted and kept cattle in New England."

All kinds of transported tradewares carry stowaway debris that may contain plant fragments. Seeds of crop plants are scarcely ever a hundred percent pure, and some weeds have certainly come in with such imports. Every load of ballast or farm implements, every cargo packed in hay or straw or even in wooden crates contains bits of plants; and it is the stiff, dry wisps bearing ripe seeds that catch on passing objects and stick most tenaciously. When the cargo is unloaded and the hold of the ship swept out, the debris ends up on some dockside dump, and any seeds in it take the first opportunity to be up and doing. It is a matter of record that many a weed new to this country was first seen by amateur botanists around wharves and shipyards, whence it spread to railroad embankments and in time has come to appear commonly along every roadside.

Sometimes the exact means of entry is not known. The handsome but pestiferous orange hawkweed, or as some call it, Indian paintbrush or king devil, was first noticed in scattered places in New England and New York State. In the 1880's, some two hundred years after Josselyn's early record, there were only small, local colonies of it; but its aggressive nature had already been noted. In the seventy years since then it has become a widespread, conspicuous part of the summer landscape along roadsides and a troublesome pest that may come to crowd the more welcome plants in poorish, dry pastures.

Field and pasture weeds have traveled along with agricultural man not merely from his ancestral home in western Europe but probably from his remote origin in Eurasia. Few botanists have concerned themselves with the classification and the complex relationships of domesticated plants, whether crop or weed. Those who have, point out that, like their American relatives, the strains of meadow plants common in European fields resemble much more closely those that grow wild in the grasslands of Asia than they do the native strains growing cheek by jowl with them in European woodlands or waste places.*

An immigrant weed has by no means an easy life. The old resident flora has been breeding and evolving for thousands of years and has produced varieties and strains enough to fill tightly all the naturally available types of living space. It is only where man breaks into the pattern and makes a new situation, albeit an old familiar one for the weed, that the newcomers can establish themselves. Gardens, barnyards, dumpheaps, roadside ditches, cultivated fields—these are the places that are constantly disturbed by man's daily round of activities and where "domestic but not domesticated" plants can thrive.

What makes a weed successful, from the weed's point of view? To make its way on open soil against the efforts of gardener or farmer, a plant must seed in abundantly, grow fast, and if possible flower and mature its seeds before it has made itself conspicuous. It is also a help to be able to get on with the business of seed production even after being uprooted and while lying prostrate in the sun.

For getting seed well dispersed into any available soil, quantity production is the thing. Dock and sheepsorrel have learned this

* See Edgar Anderson's *Plants, Man, and Life*, published by Little, Brown in 1952.

point for survival. Many weeds, like ragweed and lamb's quarters, have large numbers of small, inconspicuous flowers that rely on the universal, all-season winds for transfer of pollen. Others have the daisy type that is really a tight cluster of many minute flowers. One insect of almost any kind meandering across a few of these small bouquets can bring about cross-pollination of a large number of individual florets, each one eventually producing a seed.

Then some built-in means of transport to fresh fields is a great advantage. Consider the dandelion with its airborne fluff, and the burs and tickseeds that your long-haired pet brings home from his wanderings.

Many weeds follow a politic system of seed germination, not putting all their eggs in one basket, but sprouting a few at a time over a period of weeks or even years. If all seeds began to grow at the same time, one thorough cultivation when the seedlings are small would take care of that species for the season. Any gardener knows that it is not that simple. Each good wetting of the soil touches off a new burst of germination. Each cultivation brings more seeds to the surface where those that require light for sprouting find the conditions they need. Once a field is well seeded over, even if no more come in from the surroundings, new crops of weeds can appear in succession from the same lot of seed for many years.

In order to find out whether weed seeds really do survive as long as they seem to, several botanists have set up long-range experiments. The one that has run for the longest time was begun many years ago by Professor W. J. Beal at Michigan Agricultural College, as it then was, at East Lansing. He began his experiment in the autumn of 1879, when, in his own words: "I selected fifty freshly grown seeds from each of twenty-three different kinds of plants. Twenty such lots were prepared with the view of testing them at different times in the future. Each lot of seeds was well mixed in moderately moist sand, just as it was taken three feet below the surface, where the land had never been plowed. The seeds of each set were well mixed with the sand and placed in a pint bottle, the bottle being filled and left uncorked, and placed with the mouth slanting downwards so that the water could not accumulate about the seeds. These bottles were buried on a sandy knoll in a row running east and west and placed fifteen paces northwest from the west end of the big stone set by the class of 1873. A boulder stone, barely even with the surface soil, was set at each end of the row of bottles, which was buried about 20 inches below the surface of the ground."

Dr. Beal's plan was that he and eventually his successors would dig up a bottle every five years, plant its contents, and see what would grow. In 1920, after forty years of suspended animation, eight kinds of seeds * were still able to germinate. With the encouragement of these results it was decided to prolong the experiment by increasing the intervals between tests to ten years.

The most recent trial was made in 1950. This time, after lying buried for seventy years, three kinds of seeds resumed their active lives by sprouting forth into perfectly normal weed plants. In this endurance contest the prizes go to moth mullein, yellow dock, and evening primrose.

Other curious botanists have occasionally tried out seeds from old pressed herbarium specimens, and sometimes very old seeds have sprouted; but these were isolated trials that throw no especial light on survival in the soil under field conditions.

The weeds that are such a nuisance in cultivated fields and gardens are the plants that can pioneer onto bare, freshly exposed soil and can get along with little shelter from the elements. But the pioneer can grow only in an open, pioneer situation and cannot shoulder its way into an already settled community. Moreover, once it has become established in a place, its very presence modifies the growing conditions, shading and stabilizing the soil. With the soil anchored and with a little protection from intense heat and drying, other kinds of plants can grow among the old pioneers, which soon

* Pigweed, ragweed, black mustard, peppergrass, evening primrose, common plantain, purslane, yellow dock.

show the effect of crowding by their strenuous new competitors and begin to dwindle away. The onset of this first change marks the beginning of a succession that in New England eventually leads to forest.

Before the white man came along with his habits of stirring things up, our native knotweeds and carpetweeds had far fewer congenial places to live. Their only chance came where soil was freshly exposed by natural calamities such as spring floods or falling trees or landslides. It was a life of constant hop-skip-jump; and unless a new disturbance came along, the trees soon closed in over any gash in the green mantle, and the pioneering herbs were shaded out.

All the wide expanses of upland pasture and hayfield that are so familiar a part of New England today have been brought into existence by long generations of tree-felling. Moreover, even after the trees have long since disappeared, they will start to come back as soon as pressure is relaxed. Pastures must be grazed and fields and roadsides mowed or plowed every year to keep the brush at bay.

Any plant that is to survive in a well-kept pasture or hay meadow must be able to endure having its top cut off repeatedly. This drastic treatment calls for a special way of life; and those that live on to form the lasting population are the ones that hold their growing buds close to the ground. The prime examples of plants with this habit are the grasses. Consider the way they grow. Most kinds of grass have stems or stolons that creep along just at the soil surface, branching in all directions and sending roots down into the earth and tufts of leaves up into the air. The leaves are long and narrow, standing more or less vertically, and are renewed over a long period of time by growth from the bottom. When a grazing cow or a mower blade nips off the top ends of the leaves, the growing region is not affected, but keeps right on adding to the lower ends of the leaves. As long as there is enough herbage so that the animals do not destroy the buds themselves, the grass plants are not generally harmed.

A solid stand of turf forms a tough barrier against easy invasion by other plants, but the grass cover is rarely dense enough to exclude all comers. Many species come and go in a meadow as transients. Those that become permanent residents live as the grasses do, holding their vulnerable growing points safely below the crop-off level. The flowers and seeds of buttercups and daisies and their cohorts are borne aloft on upright stems, but if you look carefully you will find a bud enclosed in a tuft of leaves close to the ground at the base of the plant. Often the leaves form a flat rosette topping a stout taproot that goes straight down into the soil, where only drastic measures can dislodge it. The experienced digger of dandelions knows the power of their taproots to survive mutilation, form new tops, and carry on as before.

Although thousands of square miles of once-cleared fields and pastures have been abandoned to the trees, as man and his works continue to multiply, an increasingly large proportion of land is being brought under control, even where it is not strictly speaking under cultivation. The area devoted to rights of way for power lines, pipelines, and highways adds up to an impressively large total and is constantly growing. All such places need a cover of plants, if only to hold the soil in place, but they must be plants that do not interfere with access or visibility. In a naturally forested land this means chiefly keeping out the trees.

For generations man's only defense against the intrusion of forest was constant mowing or grazing. In recent years, however, an altogether new tool has been discovered in the form of the chemicals known as selective herbicides or weed killers,* which are much more deadly to broad-leaved plants of all kinds, whether weed, crop, or tree, than to grasses generally. When they are sprayed broadcast over a mixed stand of plants, only the grass and a few other resistant types survive. Moreover, the chemicals penetrate throughout the whole plant, in time killing even the farthest ramifications of the root system. This prevents formation of the suckers or root sprouts that commonly develop after the tops are cut off; and the affected plants, even large trees, are completely destroyed.

Repeated spraying with an herbicide eventually produces a stand

* The most familiar ones are known as 2,4-D and 2,4,5-T.

of almost pure grass, which is then easily kept under control by an occasional mowing. Massachusetts is currently running trials of this procedure along many of its roadsides, and the traveler can observe the results for himself, since the test areas are marked with signs. A smooth grassy roadside, green and lush through much of the year in that climate, adds a pleasingly fresh and tidy touch to the rural landscape. If there is too much of it, though, one misses the variety of colors and textures that other plants provide. The smooth sward offers no haven for wildlife, and if it is not mowed late in the season, the dead dry grass tops are a very real fire hazard.

By spraying with more discrimination one can do a rough sort of landscaping even in waste places, keeping down the cherry and poison ivy, for example, and giving the bayberry and winterberry a better chance. Gradually a pleasing composition of trees, shrubs, and open spaces can be achieved with only a modest outlay of time and labor.

Where new roads are being built, an effort is made to control from the start what grows on the broad right-of-way. There are utilitarian reasons for this, quite apart from the higher purpose of preserving as much as possible of the beauty of the land. As an aftermath of modern highway construction with giant earth-moving machinery, huge, ugly wastes of bare earth are left exposed to all the erosive forces of wind and water. Soil washed down by gullying of the slopes collects in drains and ditches and even in drifts on the pavement, whence it must be removed promptly at considerable effort and expense. The simplest way to prevent such trouble is to get the earth covered over quickly with a vigorous stand of grass.

This can be done as soon as the last grading is finished by spreading the exposed soil with a mulch of hay or straw that has grass seed mixed with it. In this way with one operation seeds are planted and the bare soil is covered with a blanket that temporarily protects it from washing. Then the mulch serves to shade the new seedlings from the hot sun until they can establish themselves firmly. The straight, slim grass leaves soon grow up through the loose mat, and in a short time the embankment is secured.

On the wide grassy verges along many of our highways and parkways there are plantings of trees and shrubs that deserve an appreciative eye. The plants used are native to the locality—maples, dogwoods, laurel, pine and the like—so that they have a strong probability of growing vigorously with no special care. They also serve to blend the roadway unobtrusively into the surrounding landscape. Plantings are made as varied and attractive as possible in order both to please the eye and to avoid the hypnotic effect that roadside monotony can have on a tired driver. In some places the roadside is a wide, closely mowed parkland dotted with large trees. Elsewhere the trees close in and the road runs as though through the woods. Then again there are drifts and clusters of shrubs that are bright in their seasons with flowers or fruits or foliage.

Where the way lies through open farmland, though, hundreds of straight young sugar maples have been planted in groups and rows along the highway edges. In time these will grow big and handsome, and we will have large-scale modern replicas of the old-fashioned tree-lined country road.

5

Running Water: Rivers and Their Valleys

IN ANY SEASON of the year New England is a well-watered land. Even after farms and forests have been moistened and lakes and bog holes filled, there is plenty of water left to run off into a maze of brooks and rivers. In fact, it would tax one's resourcefulness to remove himself an hour's foot travel from running water anywhere in the whole region.

Among the hundreds of streams there are a few little trickles in the remote hills that must go on the records as "No Name"; but all the rest have their labels, and each has its own significance. The largest rivers have kept their Indian names, or at least some European's phonetic rendering of them, and a pleasing sound they make: Housatonic, Connecticut, Kennebec, Merrimac and Penobscot; rolling Androscoggin and rippling Pemigewasset. Many of the middle-sized rivers reflect the early settlers' English origins: Westfield, Farmington, Deerfield; the White, the Taunton, the Thames (and please, it rhymes with James! *)

But it is the names of little brooks that give color to the local landscape. There are the solid old Yankee family names—Latimer, Winthrop, Wheeler, Merriam, Hale. Who do you suppose were Bill Little, Ad Chase, and Fanny Wright and more familiarly, Molly and Priscilla, that they should all have their namesake brooks?

* The New London, Connecticut, *Evening Day* not long ago ran an editorial headed, "Please do not call our river the Tems!"

Sometimes the names tell how the little streams were put to work—Oil Mill, Saw Mill, Quarry or Lime Kiln; or what manner of brook it was—Sandy or Stony, Great or Little, Grassy or Swift or Deadwater. Maybe it flowed through a Cedar Swamp or a Cranberry Meadow or Lily Pond; or was it a Salmon Hole or a Goose Pond? Once it may have been the haunt of Fox or Coon, Otter or Mink, Badger, Moose, or Bobcat. What could be the stories of the two called Blow-Me-Down and Smutty Hollow? Judging by the number of streams named for them, there was once a huge population of Beavers. And how many Roaring Brooks must there be in all of New England when just the New Hampshire part of the Connecticut River watershed has six?

Each of these streams started somewhere back in past ages as a trickle of rainwater picking its way over soil and around rocks, wandering wherever the path might lead, so long as it be downward. As the first little trickles merged into larger rivulets, gradually the growing current picked up speed and volume. Gaining strength, it began to gather up larger and larger particles of loose soil and rock from the bottom of its bed. With this the carving of a valley was under way.

A young valley, whether it is a high mountain ravine or a new gullylet in an overgrazed pasture, is always at first steep-sided and narrow, and its floor slopes sharply downhill. Water racing down over such a steep slope has a tremendous capacity for erosion and is capable of carrying a large load of sediment. In a stony land the little silt and mud that the stream bed provides is soon washed away, and most of the time the water runs transparently clear.

Slowly over the long years the stream bed deepens and the floor of the little valley is broadened. As the pitch of the river bed becomes less steep, the water current slows; and as it does so erosion tapers off until at a certain point it ceases altogether. At this moment the valley has reached its "profile of equilibrium," the delicate balance point where the stream neither erodes nor deposits, since it is carrying precisely the amount of sediment it can support, no more and no less. In this state it makes no change whatsoever in the form of its bed. Of course no stream ever stands in such perfect equilibrium with its valley for any length of time. Conditions in the natural world are not so static. Every shower brings a swelling of the stream and tips the balance toward erosion. Every run of dry weather weakens and slows the current, forcing it to lay down some of its sediment.

Stream currents and their carrying power may fluctuate widely from season to season. A brook roaring with snowmelt in April is a far different thing from the same brook in its midsummer condition of babble. In high water times, not only is there more water in the stream to take part in the carrying process, but it flows at a much faster rate. And in simple mathematical terms, just doubling the velocity of a current increases the weight of objects it can move *sixty-four-fold!* Whereas a stream flowing at one mile an hour can move gravel particles at most, the same stream racing at eleven miles an hour can roll along rocks three and a half feet in diameter. Even knowing this, it is hard to believe, after the occasional flash flood has subsided, that a mild-looking brook purling quietly along in one corner of its bed could only recently have carried away bridges, undermined buildings, and strewn a path of boulders across the nearby meadow.

Currents vary not only from time to time, but from place to place at the same time. Where a narrow channel widens and the stream spreads out into broad shallows, the current slackens and so drops the coarser part of its load. Wide places in a hill country river develop into shallow, sandy riffles, and quiet pools along a swift mountain brook become floored with clean sand and gravel. Then when the stream bed narrows again, the current funneling into it runs faster and can take on a new load of sediment.

A change of slope, too, changes the speed of a current. When a stream racing down from the hills comes out in a level valley, or where it flows into the deeper water of a lake, its speed is suddenly checked. Just where the current changes its pace, its sediments settle out layer on layer in a sloping, fan-shaped mass, or delta. The swifter the current, the coarser its first deposits, and only in the still water far out in a lake do fine silt and mud come to rest.

In the course of time the accumulating debris may spread entirely across the valley or lake bottom, filling it with a smooth, level floor. Eventually the floor may be built up far enough to stand above water level most of the time. Then the river is confined within a narrow channel that wanders over the surface that it has itself constructed. But in times of high water the river may again spread far out over the flood plain, adding to the land surface each time it does so.

A river in flood becomes heavily loaded with mud and silt. When it comes to a level flood plain, the rising water spills over its channel banks and spreads out into a wide, shallow lake. As it spreads, the drag of the land surface sharply reduces its movement, and much of the river's sediment settles out then and there. Just along the edges of the channel, where the change in flow is most abrupt, the sediments are thickest of all, piling up into ridges that may form low, natural levees. Later, when the flood subsides, the receding water may move very sluggishly and even become stagnant before it drains away, and the last drying pools and puddles serve as settling basins that trap the finest muddy clay.

Even a level flood plain is not so smooth that a river flowing over it follows a ruler-straight course; and as time goes on, the channel becomes less and less straight. At each little deflection the current washes more strongly against the outward side of the curve, cutting into the soft bank there. At the same time the inner side of the curve forms a lee shore, where the water is a little slack, so that it may even deposit a small, beach-like slope. As a result of this unequal wear, the curves in the river become more and more exaggerated, growing into deep, sac-shaped loops. In time the necks between one loop and the next become extremely narrow. Then when high water comes, the stronger flood current may cut completely through the slender isthmus to take a new and shorter course, bypassing the old river bend and leaving a crescent-shaped backwater that may remain for a long time as an oxbow pond or slough.

While the stream at the bottom of a valley is working its way downward and eventually constructing a flood plain, smaller tributaries eat into the valley sides. At the same time slopewash wears away at the sides so that they gradually become less steep. If the stream cuts through loose material, the valley may change rapidly when slumping or landslides suddenly carry down great masses of rock and soil. Solid rock walls remain precipitous for a long time, but eventually even they succumb to the forces of weathering and erosion.

As valleys widen and flatten, the divides between them become first narrower and then lower. In the end, the last, low watersheds remain only as almost imperceptible irregularities in a flat landscape.

In the course of the long ages needed to wear away mountains, the earth's surface rarely remains completely quiet. The land may become crumpled or tilted, or large areas may gently sink or rise in relation to sea level. This, of course, has its effect on the work of rivers.

If land rises, the valleys threading across it are lengthened and rivers have farther to descend before they reach the retreating sea. As a result, all the river currents flow more swiftly and begin a new cycle of vigorous erosion, cutting sharply into the floors of their old enclosing valleys. The effect shows first near the river mouths and slowly works its way upstream. Before the effect of the changed land level has reached headwaters, the lower end of a valley may have completed its new cycle of erosion and once more lie broad and flat.

Some of the rivers in the Berkshires show what may happen when a valley is rejuvenated by a rising land surface. Several million years ago the Westfield and the Deerfield, for instance, had worn their valley floors into broad, flat straths lying among low hills. Then the land rose. Since then the rivers have carved new valleys that still form steep and narrow gorges in the bedrock of the broad old valley floors, as one can see from some of the turnouts along the Mohawk Trail.

When land sinks, on the other hand, the sea rises over it, flooding up into the valleys of a rough topography or covering the whole of a flat region. In geologically recent times most of New England's river mouths have become submerged in this way. In this rough and irregular landscape, even small rivers now end grandly in large tidal estuaries that reach in between the hills, vastly extending the shoreline and carrying far inland the life of coastal waters.

New England's rivers and brooks show just about every stage of development one could wish for, though the stages are not always arranged in a logical sequence from mouth to headwaters. Wherever rocks are soft and valley-carving goes fast, the valley soon reaches maturity. The valleys of the upper Housatonic and the Hoosic Rivers and of Otter Creek, all lying along the western edge of Massachusetts and Vermont, follow a belt of soft limestone and marble. These valleys are relatively flat, and as broad as the width of the outcrop of soft rock permits. They stand in sharp contrast to other nearby valleys that cut steeply into hard granite and gneiss.

The Housatonic, for ancient geological reasons, flows from its broad upper limestone valley into a region of hard, resistant rocks where erosion is very slow. As a result, the lower valley, although chronologically at least as old, remains in a much more youthful stage. The Connecticut River behaves in much the same way. The middle part of its valley in Massachusetts and northern Connecticut is wide and well matured; but at Middletown the river leaves the soft Triassic rocks, and where its lower end flows over much more erosion-resistant terrain, its valley remains in a youthful, even gorgelike phase.

In New England young valleys often take the form of rock ravines or gorges. These are most common in the higher hills and mountains, but examples can be found almost anywhere. Many small gorges owe their existence to disruption of drainage by the glacier. In its dying stages the wasting ice directly or indirectly gave rise to thousands of ponds and lakes. Some of these were short-lived; but others existed for hundreds or thousands of years, perhaps drained through all that time by the same outlet stream. Sometimes the runoff water escaped over loose earth that eroded quickly into an open valley, but sometimes the escape channel ran over bedrock that wore away extremely slowly. In such a situation twelve thousand years of erosion have made little impression except where gravel and sand scraping along in the rapidly running water scoured out a channel, and the sidewalls of the narrow valley remain high and steep. Now that most of the lakes have drained away, some of the old stream channels are left literally high and dry on a hill-side, or cut like grooves across a hill-end spur. More often, though, there is a brook still running in the bottom of the little gorge.

The little world within a steep ravine is quite different from more exposed places nearby. High walls shut out much of the direct sunlight, and the air is cool and moist. Evaporation and even spray from the brook add to the dampness, and the surrounding walls prevent the cool, moist air from being blown away by the wind. The climate in this little pocket—the "microclimate"— is like that of less sheltered spots some distance farther to the north.

Here the vegetation, too, has a northern aspect. A stray seed sprouting into a new plant has no way of knowing whether it has fallen in an appropriate geographical location for its kind. So long as its immediate circumstances are right, the young plant continues to grow, without regard to conditions even a few inches beyond the tips of its leaves or the ends of its roots. The sheltered islet of northern climate within a steep ravine offers a poor living to lovers of warmth and sunshine, so instead of oaks and red cedars it commonly harbors more northern trees such as hemlock and yellow birch, sugar maple and beech, with a scattered undergrowth of witch-hazel, arrow-wood and hobblebush, along with the smaller woodsorrel and foamflower and the many ferns that live in the cool north woods.

Down in the brook itself and on overhanging rocks kept wet by spray or by seepage from the walls, there may be a green profusion of delicate, ferny mosses and liverworts, little plants that are not equipped to live very far from the permanent source of moisture that they find in this secluded refuge.

If the ravine is rugged enough, it may still look almost as it did in its primeval state. In inhabited regions man has been clearing and plowing, lumbering and burning for generations, so that it may be hard to know what things were like before a place was settled. But all of these operations are difficult, even impossible, in a rocky gorge; and such a place may have been little disturbed over the centuries.

Farther downstream where the valley is older, or upstream where erosion has gone faster, the sidewalls are less steep and stand farther apart, and the stream may even have begun a little flood plain at the bottom. The shallower and more open valley here offers less complete shelter from sun and wind, and the "microclimate" varies sharply from place to place. The contrast is most striking in an open ravine that runs east and west, where one slope faces north and the other south.

On the north-facing slope of the valley the sun strikes at a low angle, or direct sunlight may never penetrate at all except perhaps for a short time at midday in early summer. On a northward slope of as little as five degrees from the horizontal, the sun strikes so obliquely that the soil is as cool as it would be on a level place three hundred miles to the north. In this region of plentiful rain the soil never becomes very dry. The earth warms slowly in spring, and even exposed rocks rarely become really hot. The situation is much like that down in a steep ravine, and the vegetation, with its ferns and hemlocks, here too has a northern look.

On the opposite, south-facing slope the sun shines direct and hot. In our latitude a slope of about twenty degrees from the horizontal faces the noon June sun at a right angle. This is a truly tropical situation. For the plants growing here spring comes early and summer is hot and dry. Red cedar or hickory and oak replace the maple and hemlock, and for smaller plants there are mostly grasses, low blueberries, and sweetfern.

In the evening when the sun sets, other influences take over in the open valley. On still nights when the sky is clear, all the objects, living or not, that have been soaking up the sun's heat all day begin to radiate their warmth out into space. As a result they soon become cold, and in turn chill the layer of air that stands in contact with them. The cooled air is heavier than the overlying warmer air and tends to settle close to the ground. On an open, treeless slope, it may begin to glide downward, edging in under the warmer air above it to collect in a pool in the bottom of the valley or in any hollow or low-lying flat. Walking in the country on a still, clear summer night one can feel the coolness of the air where the road dips into a low place. Often the chill is made manifest by moisture condensed into fog patches, especially where the air is damper over a stream or marshy place; and in moonlight the course of an invisible brook is revealed by a sinuous band of mist lying over it.

In spring and fall, when night temperatures are not far above freezing, the air in low places may be just cold enough to frost-nip the tender parts of plants that are not touched on the hill-slope or on higher land. This is the reason why orchards are planted on the sides of hills, avoiding the valley floors where late frosts may damage the blossoms. Wild vegetation, too, may show the effect of frost pockets, where some of the more susceptible plants common in the general region are missing from sharply localized areas. Place names like Fog Plain or Frosty Hollow identify such cold spots for the daytime traveler.

Farther downstream where the valley levels off near sea level, or in other reaches where it has grown wide and flat, the low side-slopes have little influence. Here the valley is dominated by the broad, smooth flood plain that the river has constructed. The soil is fine-textured and silty, with no rocks at all; and the land surface shows only the gentlest undulations over a wide area. Places of this kind are commonly very fertile, and if they are not flooded too often, have a high value for agriculture.

On the lowest parts of the flood plain the natural vegetation forms a lush, grassy meadow. Standing water is never very far below the surface, and only a small rise in the river brings it lapping among the grass-stalks. Much of the year the lower leaves bear a telltale coating of mud that has not yet washed off in the rain since the last time of high water. Trees and shrubs with their deeper roots cannot grow here, and the grass remains king of the meadow.

Higher parts of the flood plain lying back from the riverbank are somewhat drier and much less often inundated. Here and on the low, ridgelike natural levees that border the channels of larger rivers grows the wet-tolerant swamp forest. This is the native home of the familiar silver maple and pin oak, along with sycamores, elms, and red maples. All of these can tolerate floodwater standing over their roots for a time every spring. They may also do well when planted in drier situations along streets and roadsides, perhaps because they are naturally adaptable and undemanding types.

Back in the swamp forests floodwaters rise and recede gently; but along the main river channel land and water are constantly shifting as new bars and islands appear and old ones are washed away in the changing currents. On a new river island the alluvial soil may perhaps be fertile, but it is exposed to the full glare of the sun; and while it is thoroughly wet in spring, and below the surface much of the time, often in summer the top layer, where seeds would germinate, is completely dried out. It takes a hardy, aggressive plant to colonize this kind of a place.

The first plants to appear on such newly deposited soil show a strong resemblance to the flora of vacant lots—another unstable, pioneer type of habitat. Both places are home for such troublesome characters as nettles and giant ragweed. Like the weeds that they are, these plants start from seed whenever conditions are right and grow fast enough to become well established before hard times come. They soon form a rank, pungent growth that anchors the soil and shelters it from the full force of the elements.

As soon as conditions are moderated by the first comers, other plants are able to seed in, and again the succession toward forest begins. The first trees that rise above the herbaceous tangle of the bottomland are black willows and poplars, often snarled up in a profusion of grape vines and poison ivy. But black willow seedlings require full sun to grow well, and cannot endure the shade cast by their parents; so as the old willows die off they are replaced by species whose young are more tolerant of shade. Gradually the nature of the woods changes as elms and maples take over, and a typical swamp forest develops.

A river island may grow rapidly, adding several yards a year to its downstream end where the water is slack. In minor floods stems and leaves dragging at the overflowing water slow it so that thick deposits of silt are built up and the island grows progressively higher. Then some year an especially strong, rushing flood cuts into the bank and sweeps away trees, bushes, soil, and whole island until the fragments run aground somewhere far downstream.

Trees growing along the river have not only water to contend with. They may also be kept back from the channel edge by the powerful thrust of winter ice. As it forms in winter and especially as it goes out in spring, the ice tears away trees and shrubs by their roots. On warm, sunny days in midwinter it expands and presses with a mighty force against the soil along the bank, pushing it up into a low, irregular ridge that rises just behind a sharp little bluff a foot or two high. Few plants can survive the shoving and churning along the channel edges, and vegetation here starts almost from scratch after an ice-bound winter.

Man, of course, may intervene even in the swampy river bottom; but he is not the only creature that modifies the course of natural events. There is also the beaver. No one seems to have made a serious study of the influence of these little animals, but indications are that it has been greater than we commonly suppose. There was a time when beavers had virtually died out in New England, except for Maine; but they have been reintroduced, and their numbers are steadily increasing.

Beavers are aquatic creatures that build themselves underwater houses. For this they may use an existing pond, but often they prefer to make their own by damming a suitable stream. To build his dam the beaver chooses a place where he can impound the most water with the least construction. Then he goes to work with logs, branches, and a judicious mixture of mud and stones. The dam may eventually be five to ten feet high and anything up to hundreds of feet long. Though it is not leak-proof even with constant repair, it can hold back a large volume of water. What the beaver has in mind, we suppose, is only to make a place to build his house; but the effect may be far-reaching.

For one thing, the impounded water with its leaky dam makes a fine regulator of stream flow as between wet and dry seasons. It serves as a flood control reservoir for excessive rains or melting snow and a source of moisture through the dry part of summer. The stream below a beaver dam flows steadily even through prolonged droughts. If there are enough beavers, their labors can have a markedly beneficial effect on the year-round water economy of an entire valley.

As time goes on the beavers' shallow pond begins to fill with dead vegetation and with silt brought in by the brook from upstream. For a long time the resident animals may keep repairing and raising their dam. But when they eventually move out or when fate catches up with them, the dam soon disintegrates and the pond gradually drains.

The situation, however, is now far different from pre-beaver. In the flooded area, which may cover many acres, large numbers of trees have been killed by standing water. Others have been gnawed down for food and for house and dam building purposes. In their place lies a large, flat expanse of silt mixed with decaying organic matter, the whole probably quite wet but not completely waterlogged. This kind of place is highly fertile and grows up to a rich grassy meadow that may persist for many years.

Beavers can be a Good Thing, or they can be an irrepressible nuisance. If one happens to want a pond where the beavers want a pond, everything is fine; and of course there are people who just like having beavers around. But when the pond floods a good hay meadow or a road, the only solution is to take the animals away. Breaking through a dam faithfully every day even for weeks on end makes absolutely no impression on the beavers, who just as faithfully repair the break and go on about their business. In western Massachusetts they are so much a factor to be reckoned with that the conservation department will on request either install beavers or live-trap and deport them from your brook.

North America had a large population of beavers before they were trapped out of most places. It has been suggested that many meadows and swamps usually regarded as old postglacial lake bottoms are really "the work of long-forgotten beavers."

6

Still Waters: Lake, Swamp and Bog

JUST BEFORE the Ice Age began there was scarcely a lake or a bog in all New England. The landscape probably looked much as the southern Appalachians do today, with hills and mountains smoothly rounded and even their summits completely covered with forest. The valleys of the many rivers and brooks had developed into a smooth and efficient drainage system, so that only in the mountains were there a few rough places left to make waterfalls, and here and there a hollow that could cradle a lake. Through long, undisturbed ages of slow erosion and weathering the land had become covered with a deep mantle of soil, formed where it lay from the underlying bedrock. The only boulders to be found lay high in the mountains or else were buried deep at the base of the soil where they had weathered out of solid mother rock. Though surface details were different then, the pattern of hills and valleys lay just about as it does today. The general scene was what a geologist's eye sees as a "maturely dissected upland with strong relief."

Then came the ice. The first wave of it came down a million years ago. There were three more waves after that, and the last one endured for a hundred thousand years. At its height the ice was thousands of feet thick and weighed trillions of tons.* It did not just lie quietly on the earth's bosom, but crept slowly over the land

* Thirty tons per square foot per thousand feet of ice thickness!

with a scraping motion like a colossal bulldozer. Since the face of New England last emerged from under this grinding burden only twelve thousand to fourteen thousand years have passed. Is it any wonder either that the ice made many changes in the land surface, or that details of the changes are still fresh and clearly visible?

The abundance of lakes, ponds, swamps, bogs, and minor wet places that now make our landscape so varied we owe to the put-and-take action of the ice sheet. While the glacier was on the move it scraped over hills, scoured deep into existing valleys, and even excavated basins in bedrock where there had been none before. In the course of its travels large quantities of rock and soil became caught up in, on, and under the ice. Eventually all of the matter gathered up was laid down again in another place. Some of it became lodged in low spots as the ice passed over. Some of it was spewed out at the glacier's edge and added to the terminal moraine and outwash sheets. Much of it came to earth only when the glacier stagnated and finally melted away. Any of these processes might produce hollows where runoff water could accumulate to make a pond or a swamp.

Most of New England's existing large lakes lie in broad, shallow basins that were scooped or quarried out by the bottom of the overriding ice sheet. Winnepesaukee in central New Hampshire and the many lakes of northern and western Maine are of this kind. There are a very few smaller lakes, such as Echo Lake in the White Mountains, that occupy basins gouged by valley glaciers before the arrival of the main continental ice sheet.

Most of our smaller lakes were formed rather by the "put" than by the "take" part of the glacier's action; and most of the putting was done after the ice had ceased to move and while it lay rotting away in ten thousand valley bottoms.

A vast amount of debris was laid down by the melting ice in the form of ground moraine or till. This kind of deposit consists largely of a jumble of rock particles of all sizes; but clay was often sorted out from other materials in the moving ice and gathered into large masses or sheets because of the strong affinity of clay for clay. A clay bed overridden by the weight of even a relatively thin layer of ice became strongly compacted and almost stonelike in texture. Any concavity among the hummocks and hollows of a moraine surface that was underlain by the hard, impervious kind of clay locally known as "hardpan" made a water-holding basin, whether it lay on high land or low; and in such situations we still find ponds and swamps, sometimes perched on high places where one would not expect to find standing water. Even a hillside slope may be marshy where either impervious clay or bedrock near the surface hinders underground drainage and so keeps the upper soil waterlogged.

Other clay beds were formed on the floors of countless ice-margin lakes. Water streaming into the lakes from the newly uncovered upland and from the still-melting ice carried a heavy load of earth debris that was laid down on the lake floor in layers of alternating coarser and finer texture as the force of the stream current varied. Much of the stratified sediment is loose and coarse and drains readily from below; but deposits that formed on the bottoms of quiet lakes may include layers of dense clay. Even a level lake floor has some undulations in its surface, and here, too, any clay-lined depression has poor bottom drainage and will hold water from one rain to the next in this climate.

Some lakes lie in old preglacial river valleys whose outlets became blocked with morainal debris, although these are rare in New England.*

Still another kind of depression that may impound water is a steep cup- or kettle-shaped hollow that has no outlet channel. These are very common in terminal moraine country, and each marks the spot where a large, detached segment of ice became buried in morainal deposits during the breakup of the glacier. Other kettles were formed where a floating ice block ran aground in an ice-margin lake and became more or less completely buried by sedi-

* The Finger Lakes of New York State are striking examples of moraine-dammed lakes, and most of the lakes in the Adirondacks apparently originated in this way.

ments washing in around it. In either case, when the ice finally melted out from under the debris, the enclosing earth caved in over the spot, leaving a sharply-defined hollow with steep, smooth sides formed by slumping of the soft earth. Sometimes kettles appear in a row where a long arm of disintegrating ice became scored with a series of crosswise crevasses that filled with debris. When the broken ice finally melted, the crevasse fillings remained as partitions between one kettle hollow and the next.

Most kettles appear in loose sandy or gravelly deposits and are dry on the bottom. If the kettle is deep enough to reach down to underground water, or if drainage is impeded by rock or hardpan, the hollow is boggy or may even hold a stagnant pond.

On flat places ready drainage is sometimes obstructed by the long, winding ridges of eskers. These remarkable constructions, steep-sided and flat-topped, appear to consist of sediments laid down by streams that flowed in tunnels beneath the wasting ice. They are most common on the floors of low, flat valleys such as that of the Kennebec River in Maine. They impede the free run-off of surface water and their winding curves are especially likely to enclose shallow swamps.

A few ponds and sloughs owe their existence not to the glacier but to recent river erosion. These are the curving oxbows cut off where a river once in flood time took a short cut across a meander neck. Perhaps for the sake of completeness one should mention also beaver ponds.

Whatever its origin, a lake is geologically speaking a short-lived affair. As soon as it appears, natural processes set about obliterating it by simultaneous draining and filling. The draining is brought about by the surface stream through which water escapes from the lake basin. As the outlet stream gradually deepens its bed, the level of the spillway from lake into stream is correspondingly lowered, and the lake water becomes shallower and shallower. At the same time the bottom of the lake basin fills with debris brought down by streams flowing into it. Any vegetation growing in the lake contributes to the filling process by dragging at the water currents, slowing them so that they unload more and more of their sediments. When the vegetation at last dies and decays, its own remains are added to the lake bottom accumulation. It is not very long before the one-time lake becomes only a flat place along the river valley.

The flat intervale meadows of the White Mountains are examples of old filled-in lakes that originally lay in hollows on the irregular surface of the ground moraine after the glacier first disappeared. The lakes have since become filled and drained, and their sites now provide the little level, farmable land in that rugged countryside.

While a lake lasts, the vegetation in and around it is arranged in a series of zones, according to the depth of the water. As the lake becomes smaller and shallower, the vegetation zones gradually move toward the center of the remaining water.

Out in the middle where the lake is deepest, only free-floating plants can grow. Rafts of duckweed may drift about on the surface, collecting now here and now there, wherever winds and currents take them. Tangles of hornwort or bladderwort float in the upper layers of the water, completely submerged except for a brief time when their flowers are pushed up into the air where insects can reach them to bring about cross-pollination.

Toward shore where the water is shallower, plants appear that are rooted in the silt or mud of the lake bottom. If the water is very clear, rooted plants can grow to depths of about twenty feet; but if the water is in the least colored or cloudy, light penetration is greatly reduced, and even a few feet down it may be too dark for any green things at all to live. Of the rooted plants, those that push farthest out from shore live completely submerged, their soft, feathery leaves never reaching the surface. The very names of these plants often suggest their appearance: *Myriophyllum*, the water milfoil; *Hippuris*, the mare's tail; *Proserpinaca*, the mermaid weed. Water only six or eight feet deep is shallow enough for plants with leaves that float on the surface, attached by long, flexible stalks to roots and rhizomes anchored in the mud below. Most familiar of these are the various waterlilies, whose names have a classical ring: *Naias, Nymphaea, Castalia.*

For a time floaters and submerged plants may grow intermingled; but as lily pads cover the surface more and more thickly and the underwater light becomes dimmer, the submerged plants begin to weaken and die, adding their remains to the silt and mud that accumulate in the still water around them. The succulent tissues of gone-by waterlilies, too, add substantially to the bottom debris.

In the airless underwater mud, decay is extremely slow, and the pond bottom may build up rapidly with dead vegetation. As it does so the shallower water offers a foothold to other, stouter plants that also live partly submerged but push their leaves up above the surface of the water. Of the many such that are found in New England, bullrushes can grow in as much as six feet of water; pickerelweed requires somewhat less; and cattails usually prefer it only a few inches deep. In the shallows there may also be stands of reeds, wild rice, arrowhead, sweet flag or water plantain, or mixtures of any or all of these.

The tall, upright foliage of the newcomers casts an increasingly dense shade on leaves floating flat on the water's surface. Lilies escape by migrating out to the edge of deep water. If the bottom drops abruptly they may form a narrow band near the shore; or a shallow pond may become entirely carpeted over with them—an enchanting sight when the pale flowers are spread on a background of black water in some quiet woodland pool.

The tall, narrow leaves and brown seed spikes of a cattail swamp are a common feature of our landscape, green through the summer, turning tawny golden in autumn, and softening to tan through the winter. They may cover large areas in quiet backwaters along a slow river or at the end of a lake; or they may form a hedgelike strip in a shallow ditch by the roadside. Wherever they grow, cattails are a sign of permanent shallow water.

Swamp vegetation in its turn contributes to the raising of the soil level; and here and there appear spots where the water is too shallow even for cattails and the like. As usual there are other plants ready to move into any available place where conditions suit them; and the sunny expanse of swamp, by now almost solid but still very wet, gradually transforms into a sedge meadow.

Sedges, to a casual glance, look much like their relatives, the grasses; but most sedges are a more yellowish-green and shinier, and they generally live in wetter or more difficult places. On the meadow they grow in a characteristic way to form dense tussocks that may stand like islands several inches above their immediate surroundings. All kinds of debris becomes caught around them; and as their old tissues die and new ones grow above them, the sedge clumps develop into small hummocks that are scattered irregularly over the swampy meadow. Slight differences in height above the soil make a large difference in wetness for other small plants; and these may grow in a crazy quilt patchwork, each one perched at its own preferred level among the sedge tussocks. Here and there the meadow may be brightened with the color of golden marsh marigolds or blue flags in their seasons of bloom.

When the meadow reaches the point where its surface is above water nearly all summer, small trees and shrubs begin to appear. Alder and buttonbush are very tolerant of wet feet, and along with shrubby willows and sweet pepperbush they may form a dense, head-high thicket just at the edge of the meadow. The countless leaves of this large mass of herbage evaporate large quantities of water that they draw up through the roots of the plants from the earth below. As a result, the soil may become drier than its height alone would lead one to believe.

When the soil is no longer waterlogged, except perhaps in winter and early spring, air can filter into it, allowing the accumulated organic matter to decay into a fine humus, which serves to enrich the silty mineral soil mixed with it. Circumstances now resemble those on the higher flood plain of a river; and the erstwhile meadow gives way to a woodland where red maple, elm, yellow birch and their usual companions predominate, as in a flood plain swamp forest. As the land underfoot in time becomes still higher and drier, these in their turn give way to the truly climax forest of the region.

The changing world of a filling lake bed is one of the classic places to study the process of plant succession—the constantly varying procession of plants that grow in a given place over a long period of time. Whenever a new pond appears, or an open field is abandoned or a forest destroyed by fire, only a few kinds of plants find the new conditions to their liking, even in the absence of competing plants. Those few move into the bare place as pioneers. But the pioneers shade the soil, and their dead remains enrich it; and soon other plants appear among them and gradually shoulder them out. The newcomers make further changes in the living conditions that encourage the growth of still other plants, and so on down the line. The succession at last comes to an end when mature plants perpetuate the conditions that favor the growth of their own offspring under the local circumstances.

The group of plants that finally persists is called the "climax" vegetation. The climax for a given place is governed in the last resort by soil and climate. So long as the climate remains constant, the climax, once established, will tend to perpetuate itself. If a disturbance such as fire, flood, or hurricane changes the growing conditions, the succession may be set back a stage or two; but ultimately the pressure of events takes it on to the climax stage again.

Successional changes in a freshwater lake can be observed over a period of years when a new dam is built or when a new lake is formed as the result, for example, of an earthquake or a landslide; but sometimes the complete story can be reconstructed from stages that appear all at one time around the edge of an existing lake. Since the fate of a lake is to become progressively smaller and shallower until eventually it is extinguished, the earliest phase appears in deep water, with succeeding ones laid out in order through progressively shallower water to the water's edge and on across to dry upland.

Not every lake shows the cattail, sedge meadow, swamp forest succession. Sometimes the open water is fringed with a floating mat of low, shrubby plants entangled with bogmoss. Around this lies a zone of heathlike evergreen shrubs, backed in turn by the spires of spruce and tamarack. In northern regions * the trees merge imperceptibly into the surrounding evergreen forest; but farther south such a bog with its strange plants appears like an island sharply marked off from the broad-leaved forest around it.

Now, the terms "bog," "swamp," and "marsh" are used more or less interchangeably in common parlance. But students of vegetation have come to use the word "swamp" for the cattail-to-maple woods places and reserve "bog" for the moss, heath shrub, spruce situations. One would have a hard time composing a scientifically precise definition of a bog that would suit all botanists, but probably all of them would point to the same kinds of places as examples of bogs.

Bogs are most commonly found in regions with a decidedly cool, moist climate. Where the general climate is not especially favorable, a bog may still appear in the kind of low-lying, sheltered spot that forms a frost pocket, where the "microclimate" is distinctly cooler than that of the surrounding country.

The kind of hollow where a bog is likely to develop is poorly drained. The water standing in it has very little natural circulation, and the dense plant growth restricts what movement there is. With no currents strong enough to carry it away, dead organic matter accumulates where it falls. Decay is extremely slow, since the water is cold and contains very little oxygen. It is also strongly acid, low in mineral content, and has a characteristic brownish color rather like bad coffee. It is no cause for surprise that a unique group of plants inhabits such places.

A venture into the depths of a bog on foot is a journey into a strange and unfamiliar world. The edge of the bog, where solid earth falls away, is often marked by a sort of moat several feet wide of more or less open water. This can probably be crossed on a fallen tree or jumped at a narrow place. Then comes a bit of hard going through a dense growth of trees and shrubs that will discourage the faint-hearted or delicately clad. Coming out on the low, shrubby expanse of the "quaking bog" itself one has an inkling of

* In the far North a mossy bog strewn with sedge tussocks and low shrubs is called by the Indian name of "muskeg."

how Alice felt in Wonderland. Standing still, you feel the "ground" undulate beneath your feet whenever someone walks past a few yards away. A gentle bouncing up and down sets the trees wobbling for several yards around. When one person wants to take a picture, everyone must stand still so that no footsteps will jiggle the camera. If you stand in one spot for several minutes, the water almost imperceptibly creeps over your feet and up your shins as the floating mat slowly settles. As you move around you learn to avoid anything that looks like a path and walk instead on the tops of the most solid-looking bushes. Nearly everyone steps on a thin place eventually and goes in above the knee. Happily, although the mat is flexible, it is tough, and the holes in it are small; so there is little danger of going all the way through to the lake bottom.

Animal life is rather sparse in a bog. Insects are abundant, and there may be small birds that live in the trees and brush. Deer may wander in to browse, although one wonders how they manage on the loose-textured mat with their small hooves. But compared to the teeming life of a marsh, the bog is a quiet, vacant place.

Like other lakes, a bog lake fills from the edges inward. Any open water that remains in the center is bordered by a loose network of floating plants, chiefly sedges and waterwillow, that spread by means of arching stems or stolons. Since the stems grow out in all directions, they become thoroughly intertwined to form a strong, resilient mat of low-growing twigs and branches. On top of this and in the interstices grows a luxuriant mass of bogmoss or sphagnum. Just a few feet back from the edge the floating mat may be thick and strong enough to support the weight of several people, if they take care where they step and do not all crowd into one place.

On the tangled mat grows a characteristic assemblage of plants, including the bizarre sundews and pitcher plants that catch and digest insects. These they use to augment the nitrogen that is deficient in bog water and not otherwise available to small plants whose roots do not reach down through the water and peat to the underlying mineral soil.

Many other bog plants are evergreen members of the heath family, among them cranberries, leatherleaf, Labrador tea, andromeda and the pale laurel. These and a few other unrelated plants that grow with them all have in common the thick, leathery leaves that apparently serve to reduce evaporation of water. This would presumably adapt them to living in dry places. Now, a bog looks like anything but a dry place, and many attempts have been made to explain why drought-resistant plants should grow in standing water. One explanation offered is that a bog is "physiologically dry" because the water there contains so much dissolved matter that plant roots have difficulty absorbing water from it. This produces a result something like "burning" a lawn or a garden with too much fertilizer. Another thought is that since bog plants grow in a shallow "soil," the size of their root systems and hence their ability to absorb water is limited by the nearness of standing water, just as much as though the plants were growing on rock. Another suggestion is that lack of water is a problem chiefly in winter when the bog may be frozen for long periods of time or when in any case the temperature around the roots is so low that they cannot take in water fast enough to replace what evaporates into the warmer air around the evergreen leaves.

Most of this is sheer armchair speculation. The little really experimental investigating that has been done suggests that the story may be discouragingly complex; and the relation of bog plants to bog conditions is not at present a popular field of research. On the other hand, it might conceivably turn out that bog plants are simply those that can make do in a cold, sterile habitat, whether wet or dry.

Farther back from the open water, where the bog mat is older and thicker, small elevations begin to push up above the general sogginess. On the relatively dry hummocks trees begin to appear, most commonly black spruce and tamarack, along with occasional red maples, birches, and mountain ash. In more southern bogs white cedar and rhododendron are common.

The last two species are abundant in the bogs of southeastern Connecticut and Rhode Island but nowhere else in New England.

The nearest place they are found is New Jersey, and they are common on the coastal plain southward from there. This leaves a small island of them separated from the great mass of their relatives to the south by a gap of a hundred and fifty miles. One may wonder why they are found in New England at all.

The available evidence leads us to believe that the white cedar and some of its companions migrated northward long ago in the wake of the retreating glacier, following a broad coastal plain that has since become submerged beneath the ocean. Conditions that allow the growth of these plants do not extend beyond the southeastern part of New England; and since their old southward connecting path has disappeared, they are caught in a small, isolated pocket.

There are various reasons why this explanation is scientifically acceptable. For one thing, there are other, strictly geological grounds for assuming a now-sunken coastal plain off our shores. Also, many other kinds of plants are known that now grow only in widely separated places. In nearly all such cases there is reason to believe that the plants formerly grew over a much wider area that included both the isolated places where they are still found and the intervening territory where they are not. Sometimes the plants themselves are found as fossils in various localities where they are now extinct. In a number of cases it is clear that a geographical range that was once large and continuous has been interrupted by the rising of mountains or the intrusion of oceans.

Another botanical puzzle is the existence of northern plants in isolated bogs far south of their general range. How did they get there? Here again it is believed that the plants formerly grew throughout a large region where they no longer exist. When the Ice Age came on, they were crowded southward from their former homes into such tolerable spots as they could find among the plants already in residence. Then as the ice retreated, the northern plants moved back again; but a few individuals and their progeny hung on in small localities along the way that especially suited them and where they could hold their own against the pressure of returning temperate-climate plants.

There are those who believe that present conditions can account quite adequately for the nature of bog vegetation without calling in any glacial relict theories. We have already mentioned that bogs commonly lie in frost pockets where temperatures are low and the growing season short. Once the water in a bog is frozen, the thick mat of organic matter serves as excellent insulation to keep it so. As an example of this, there is the case of a huge bog at Harrington, Maine. A man taking borings of the peat there one year late in April encountered a layer three inches thick that was still frozen so hard that he had to use an iron bar to break through it. Other instances are reported of frost lingering in the bottom of bogs as late as early July. There is evidence, though, that where northern and southern plants are growing together, the southerners are getting the upper hand. This would indicate that present conditions in those places are not such as to favor the survival of cold climate plants.

Bog vegetation is not always easily displaced by other plants. Where cool, wet conditions encourage the luxuriant growth of sphagnum moss, it can increase so rapidly that it spreads out and smothers the adjacent forest. The moss and its accompanying bog plants may continue to spread until eventually in one way or another changing conditions restrain them. This may come about as the result of either improved drainage or increased evaporation from the surface. Manipulation of river flow for logging purposes may do the former; increased exposure to wind after logging or fire may accomplish the latter.

In a truly oceanic climate, very cool and very wet, bogs may form on level or even slightly sloping places, so long as there is a plentiful supply of moisture available from below. In such places sphagnum moss may grow so profusely that it builds up a great, wet cushion that may be higher in the center than at the edges by anything from a few inches to several feet. This is like the raised bog or Hochmoor of northern Europe and Scandinavia. In New

England raised bogs are found only in a narrow strip paralleling the coast of eastern Maine, mostly beyond Penobscot Bay. Elsewhere, though the climate may seem damp enough, evaporation is too great to allow a bog to grow upward from a flat surface.

As time goes on, the floating vegetation of a bog pond encroaches further and further into the open water, finally covering it over completely. Meanwhile, as old plants die, their remains are pushed down into the water by the weight of new growth above them. Some of the fragments come loose from the underside of the mat and fall through the water to the bottom. Gradually the original pond becomes filled with almost pure organic peat.

If the one-time bog hollow has any drainage outlet at all, it may in its last stages of filling turn into a shallow swamp where the water flows in and out again rather freely. Then the spruce of a typical bog gives way to the red maples and elms of a swamp forest. If drainage of the bog improves suddenly for any reason, the trees that have been struggling along in a stunted condition show a great spurt of growth and develop rapidly into a normally vigorous forest.

In time the upper layers of the old peat bed dry out enough to decay into humus and become mixed with mineral soil so that they will support the climax forest of the region. Then the bog is completely obliterated and its existence may never be suspected until heavy buildings or roads constructed on top of it begin to settle as their weight compacts the resilient peat.

In New England's irregular topography a small stretch of country may have many different kinds of wet and dry places. These are not always situated where one might expect to find them. There may be lakes on hilltops, marshes on the slopes, and dry sand plains on the valley floors. Nearly all such anomalies are the work of the glacier. They are all transient, and a few aeons will see the end of them all.

7

The Shore: Headland, Dune and Salt Marsh

A MOTORIST FOLLOWING the not especially straight route of U. S. Highway 1 from Greenwich, Connecticut, to Eastport, Maine, travels a little more than five hundred miles. If he did the journey on foot following high tide line around the headlands and skirting every bay and estuary, he would travel many times that distance. A trip around every offshore island would add still more hundreds of miles. On this long walk the footing would vary from precipitous rock faces and sandy cliffs through every kind of ledge and boulder, narrow shingle and wide sandy beach, to flat marsh where it is hard to say just where sea ends and land begins.

On the other hand, if our traveler went up in a space ship and cruised at a height where details disappeared but where he could see the lay of the entire New England coastline, it would stretch below him as a long, loosely S-shaped curve. Along this line the land surface, scored with hills and valleys but in the larger view essentially flat, slopes gently downward toward the southeast and dips below the level of the sea.

The waters of the ocean today wash much farther inland over the face of New England than they did in times that are geologically speaking only yesterday. At the height of the Ice Age so much of earth's water was locked up in glacial ice lying on top of the land that oceans the world over stood markedly lower than they do now. While the waters were withdrawn, the broad coastal plain that lies off our shore was exposed as dry land, perhaps as far out as the very edge of the continental shelf. While it lasted this coastal plain must have looked much as the wet, sandy lowlands of tidewater regions of Virginia and New Jersey do today.

When at last the grand thawing removed the ice cover from New England and other northern lands, such large quantities of water flowed back into the ocean basins that seas all over the world returned once more to higher levels, and the waters again flooded far inland over the coastal plain that had only recently been uncovered off New England.

While seas were rising with the influx of glacial meltwater, the crust of the earth stirred slightly, as it does from time to time; and as the end result of several fairly minor changes of level, the entire New England coast became tilted downward toward the northeast. The Connecticut shore has sunk relatively little; but the bottom of the Gulf of Maine lies at least twelve hundred feet lower than it did not so long ago as geological time is reckoned; and along the present Maine coast the sea has flooded far up into erstwhile valleys and breaks against what once were upland hillsides.

There is evidence that at the present time sea level is gradually rising all along our Atlantic seaboard. But careful study of the shore and calculations from rates of change where the waves are now eating away at cliffs and dunes along the New England coast lead to the conclusion that the sea has stood at approximately its present position here for some four or five thousand years. This is a long enough time to make great changes in loose earth, but scarcely enough to scratch the surface of hard, crystalline bedrock.

The entire shore from New York to Maine and even beyond may appear as a single region to a geologist; but to the ordinary traveler's eye it changes markedly from place to place. Beginning at the top there is the ragged, rocky coast of Maine, more or less similar from Eastport to Portland. In most of Maine the primordial mountains were never entirely worn away, and the land is still hilly. The shoreline here lies east and west, crosswise to the rough grain of the land; and on this sunken coast, long fingers of the sea reach up into narrow, parallel valleys. Where the land slopes off below the water, former hills are left as rocky islands and treacherous submerged ledges standing out to sea. The result is a fantastically ragged and picturesque reach of coastline.

The rocky skeleton of this land is never far hidden. Its original soil was largely removed by the glacier, and the ground moraine that the ice left as it melted was quickly washed away by waves of the returning sea, exposing bedrock all along the shore. In the bare rocks of Mount Desert Island, midway along the Maine coast, records of the history of the land are spread for an attentive eye to read.

Standing at the water's edge there, a spectator is awestruck by the power of pounding waves to tear at the rocks and hurl them around at the tide line. But let him take a longer view and notice the larger contours of the earth. At a distance even the tallest rocky cliffs scarcely show on the skyline profile of the land as it dips below the water. There is scarcely any sign of the wavecut terrace that would shelve out just below the waterline if the sea had eaten very far into the cliffs. Tremendous as the pounding of storm waves may be, they have as yet scarcely nicked into the mass of exposed rock.

The sea cliffs, however, are impressive enough compared to human size. It is their steepness that makes them so striking. But the hillsides out of which they were cut were steep to begin with, especially on the south or seaward side. Inland hills nearby but far removed from the influence of sea waves are shaped just like the coastal promontories, with long gentle north slopes and precipitous south faces. Hills and headlands alike are formed from rocks that break easily along natural vertical planes of weakness. Since the rocks do not lie horizontal but slope downward toward the north, it is the southern exposures that form steep cliff faces. If the loose rocks and soil at their bases were taken away, the hills would make dramatic looking declivities much like the ocean-facing headlands.

More widespread than the work of the sea are marks of the

glacier's passing. Away from the cliffs, in lower places where even a gentle ocean swell washes far up over the sloping rocks, a sharp eye can find ice-carved grooves and scratches reaching down to just above high tide line.

Some of the most spectacular effects of ice on native rock can be seen in the hills of Mount Desert Island. These solid granite mountains, rising fifteen hundred feet almost directly from the water's edge, were shaped from a resistant monadnock ridge that stretches for eleven miles east and west across the island. The ridge was probably already nicked with shallow north-south notches when the advancing glacier came broadside upon it. As the ice rose and in its stately way poured through and around the gaps and eventually over the crest of the ridge, it scraped the hilltops into their present rounded contours and scoured the sides and floors of the valleys until the onetime shallow notches were dug out into deep, U-shaped troughs. Now most of the rock-floored troughs contain freshwater lakes. One of them is open to the sea, and Somes Sound lying in it can just about qualify by virtue of its glacial origin as New England's only true fjord.

An exposed rocky shore like the Maine coast has little sand and gravel for waves and currents to form into sand bars and beaches. Tucked back between the headlands there are little pocket beaches where the breaking waves drop what loose matter they carry, and there are occasional bars and sandspits. But bare rock washed by cold, clear water dominates the scene for mile upon rugged mile.

Then south of Portland and down as far as the rocky outcrop of Cape Ann the waves wash in over broad expanses of sand. Hills fall back inland, and only an occasional lumpy ridge reaches out to the ocean. Here is a little bit of true coastal plain, a flat expanse of sand and silt that rivers once brought down from the hills and spread in the shallow edge of the sea. A minor and local bulge in the earth's crust then lifted the little plain above water, and now it stands as a flat shelf fringing the bases of hills and islands that once were washed directly by the ocean. In a few places the plain reaches as much as fifty miles inland. More commonly it is much narrower.

The large curve of Massachusetts Bay has a different history. The Boston Basin has been a low-lying area for a long time, even in geological terms. Far back in the Carboniferous age, when the great coal beds all over the world were laid down, the basin became floored over with sediments dropped in a shallow sea. Though the floor has been lifted up and warped a little in the ages since, it still lies fairly level and undisturbed. Its soft rocks were long ago worn into a broad lowland; and here the melting glacier left a cover of ground moraine and a great array of drumlins.

Some of these whaleback mounds stand as islands in Boston Bay; some form inland hills; and many more are strewn along just at the water's edge. Where waves have access to glacial moraines and drumlins they rapidly rework the loose deposits into forms of their own design.

When water reaches the base of a drumlin its first effect is to eat into its seaward edge. This forms a cliff that becomes higher and higher as it progresses into the sloping hillside. After the cliff face has reached the crest of the hill, its height gradually diminishes until the last remnant of the hill has been washed away and the drumlin is no more.

In the course of several thousand years many drumlins must have been taken by the sea. A number of islands have disappeared from Boston Harbor just in the three hundred years of historic time. The twelve-acre Nixes Island, which was granted to John Galop in 1636, remains today only as a bouldery shoal that is of little use to the present heirs.

The earth that is washed away from the shore is not just carried out to sea and lost, but rather is added to the loose material constantly moving along in the shore currents. As waves break against the shore, they carry up with them whatever sand and stones they have already picked up; and as each wave retreats it leaves behind part of its original load, dragging the rest away along with a certain amount of new material loosened from the shore. Heavier boulders dislodged from the upland may lie where they fall, scarcely rocked by pounding water, and gradually build up a natural sea wall that protects the slope above them from further cutting. Smaller stones and especially clay and silt become added to the debris that is forever shifting along the shore.

Water movement along the shore has several driving forces. Rivers flowing out from the land provide a certain momentum. The rise and fall of tides has its effect, although this is small except where the tidewaters race in and out through narrow passages. But the chief force that keeps the longshore currents moving is that of waves striking the land obliquely.

The breaking of waves is a mechanical process governed by precise mathematical relationships. A swell of a given height and breadth will rise to a crest and tumble over itself when it rolls into water of exactly a certain depth. Since the depth contours of a sea or lake bottom never lie quite parallel to the waterline, waves always break at more or less of an angle to the shore. As a wave slides back down from the beach, however, it does not retrace its old oblique path but moves out more directly seaward. As a result, every sand particle and every piece of driftwood or other flotsam carried in on the breakers moves erratically along the shore—obliquely in and directly out—until eventually it is tossed far enough up on the beach to become stranded.

On a gently sloping shore the line of breakers may be far out from the water's edge. The breaking of a wave checks its speed sharply, making it drop some of its sediment. This raises the bottom, and the shallower water then intensifies the breaking action of the waves. Gradually the bottom along the breaker line builds up into an offshore bar. In time this may rise above all but the highest reach of the water, and in quiet weather stand exposed as a barrier beach.

If the bar touches land at one end, it forms a spit, which tends to keep growing at its outer, free end. If the spit meets deeper water where a river flows out from land, or if it encounters a current moving head-on toward it around a point or headland, its growing end may curve around into various kinds of hooks or loops. The change in form of the spit then changes the pattern of longshore currents, and this in turn reacts on the growth of the spit. At any time during this complex interplay of forces the huge waves of a severe storm may intervene to reshape the spits and bars and drastically change the course of events.

On irregular shores, waves washing into embayments may construct bars or beaches at baymouth, midbay or bayhead, according to the position of the breaker line. Islands fringing the coast may become tied to the shore or to each other by bars or spits to produce forms known as tombolos. In the Boston Bay area Nantasket Beach and the Winthrop region show beautiful examples of complex tombolos made by the tying together of several drumlin islands.

Cape Cod and the nearby islands owe their existence to the ice sheet. At the height of the glacier's last advance, the southern edge of the ice stood in three great lobes that extended into Cap Cod Bay, Buzzard's Bay and Long Island Sound. For thousands of years the ice front remained virtually stationary in this position. Through all this time rock debris of all kinds released at the melting ice front accumulated to form a terminal moraine, an irregular lumpy ridge that lay in festoons fringing the great lobes of ice. Parts of the morainal ridge remain today, forming the hilly parts of the Plymouth region, the Cape, the Elizabeth Islands, the Rhode Island shore, Fisher's Island, and Orient Point on Long Island. An earlier ice front in the same way left us Nantucket, Martha's Vineyard, Block Island, and Montauk Point.

The moraines are a sterile mishmash of sand, clay, and boulders heaped into chaotic swells and hummocks, with hollows between that hold countless ponds and swamps. Bordering the moraines on their seaward sides are broad, gently sloping outwash plains. These are composites of many deltas made by torrential streams of meltwater and rain that poured down over the glacier and its moraines, finally spreading their sediments where they flowed into the edge of the sea, or perhaps onto the level coastal plain while it stood above the water.

Cape Cod has fascinated many people for many reasons. Among these are various geologists, who have made use of all that is known of both the constructive and the erosive action of the sea to reconstruct the life history of the Cape. The "wrist" and "hand" of this beckoning sandy arm consist largely of beaches and spits and their enclosed marshes, all clearly the products of wave action. If these were all stripped away, the Cape would end in a row of sea cliffs that rise in a long scarp about a mile north of a line drawn from Truro to Highland Light. The cliffs were originally formed by waves cutting into the north sides of morainal hills and delta plains. Now they have been abandoned by the sea, and old beaches that formed long ago at their feet protect them from further cutting. All the sandy expanse of the Province Lands to the north and west of this line has since been constructed by shore currents and then modeled by the wind.

South of Truro the "forearm" has in the course of time become narrower, especially on the eastern or back side exposed to the open Atlantic. Here the pounding waves are steadily eating into the loosely-packed soil of the cliffs. All irregularities in the shoreline have been worn away, and cliffs and beaches stretch for miles in a long, straight line. Complex calculations lead us to conclude that the sea has been working at the outer Cape for some three or four thousand years and that, other things being equal, another four or five thousand will see this lovely sandy waste completely washed into the ocean.

Along the south shore of the "upper arm" and more clearly on Nantucket and the Vineyard, waves have made less drastic changes in the shoreline. Here they have merely cliffed the ends of the fringed lobes of the outwash plain and connected long stretches of the cliffs with baymouth bars and beaches. Behind the outer beaches lie great marshy ponds, some open to the sea and salty, others completely closed and more or less fresh.

On the Rhode Island shore the more northerly of the terminal moraines intersects the sea at Watch Hill. This region has been called "one of the finest examples of glacial dumping ground in the entire east." It has no outcrops of bedrock, but is a mass of knobs and irregular ridges spotted with undrained kettle hollows and a fine display of glacier-dropped boulders. There is one especially appealing little kame shaped like a sugarloaf and only about ten feet high, just beside the road from Westerly to Watch Hill, still standing where the ice set it down in a neat little pile.

Between Westerly and the town of Narragansett the main highway first crosses the moraine and then skirts its southern edge, offering a fine view of a narrow outwash plain sloping down to the sea, with salt ponds bordered by low ridges crossing the plain. Parts of the plain are fertile enough for good potato farming, although only small areas now stand above water. Beyond, fringing the open ocean, lies a long strip of beach and dune, thickly studded with summer homes.

This bit of unprotected coast suffers most severely from the direct onslaught of occasional wandering hurricanes. Exceptionally high tides and storm-driven waves may wash clean over the outer beach ridge, flattening the dunes and depositing dune sand along with cottages and their contents in the salt ponds behind. In some places the sandspits and beaches change rapidly enough to instill caution in builders of summer homes. But the beaches at Weekapaug and Misquamicut had not been much changed since the great storm of 1815 when the violent hurricane of 1938 came along and washed away most of a century's accumulation of cottages.

The indentation of Narragansett Bay could be likened to a Boston Bay without the spattering of drumlins. Here, too, the sea reaches far up into an old lowland that owes its existence to a surface exposure of weak, easily eroded rock. But in this region the rocks have been strongly folded lengthwise. This is strikingly clear as one drives along a level, north-south ridge crest, then turns off onto a switchback east-west road. Any map shows, too, the resulting long, narrow forms of the bay's islands: Conanicut, Prudence, Patience and Hope, and Rhode Island itself with its ancient seafaring town of Newport.

The coast of Connecticut is perhaps least spectacular, but fully as charming as any in New England. It has its quota of rocky headlands, beaches, bays and salt marshes. Even this shore protected by

Long Island has its problems of erosion. Every year a certain amount of valuable waterfront real estate slips down the bank onto the beaches or washes into the Sound. The moderate waves of summertime are relatively innocuous. It is the exceptionally violent winter storms with high tides and driving waves that make sudden great changes. Sea walls to be effective must be massive and high enough to serve as shields against the most tremendous storms.

Along all these varied types of sea-front, the vegetation grows down as close to the water as conditions allow. Shore life of all kinds is subject to almost constant wind and to various degrees of salt spray and wave and tidal movement. These factors have a strong selective action on plants that grow along the coast, and the susceptible are vigorously weeded out.

The wind imposes several kinds of strain on plants. Mere physical whipping about can injure tender new growth so badly that branches never succeed in establishing themselves higher than the level of protecting slopes or walls or other vegetation. Even more restrictive is the wind's drying effect. Plants that can survive these hazards may be done in by windblown salt spray.

In sheltered spots it is soil that determines how close to the water the inland forest can grow. On rocky shores loose soil is rapidly washed away, and near the waterline only little crevice pockets remain where small plants like grasses and sedges, or stunted trees and shrubs, can find an anchorage. Earthen shores offer good footing for plants, although wave erosion keeps eating into the available space; and only where such places are exposed to strong winds does their seaside vegetation differ much from that of the nearby inland.

In places where the sea is making new land by depositing sand or mud, the chief problems for plants are salty water, intense sun, and unstable underpinnings. In the quiet water of sheltered bays and ponds the finest silt and mud are deposited, giving rise to marshes. In moving water the deposits are of coarser sand and gravel that is perennially worked and reworked by the action of waves and currents.

A shifting sandy expanse washed twice every day by the waves offers a poor foothold for any kind of fixed creature, and the lower beach as far up as the waves of ordinary summer storms reach is practically destitute of plants.

Above this lies a strip of "drift beach" that is wave-washed by the severest winter storms but remains quiet all summer. Here grow annual plants that live out their lives from seed to seed in a single season and are indifferent to being uprooted or torn apart during the winter. Even on an undisturbed beach not many plants can survive the intense heat and glare and salt spray, and vegetation is accordingly sparse. Only a few rugged species can live there, such as various kinds of sea-rocket, cocklebur, saltwort, orach and spurge.

On the upper beach, above the reach of even the highest waves, perennials can safely establish themselves. This is the home of the beach pea, the fleshy little sandwort, and the dusty miller that is a relative of the sagebrush of our western plains. But the beach plant par excellence is the sand reed or beach grass—*Ammophila*, "the sand lover." This coarse, harsh, sawtooth-leaved plant may dominate the landscape for miles along the shore. As it grows it sends out creeping, horizontal stems that give rise at the tips to tufts of new leaves and clusters of long, fibrous roots. If shifting sand rises around the plant, it grows fast enough to keep new leaves pushing out into the air. A hill of sand covered with a good growth of beach grass becomes securely stabilized, and only the most strenuous attacks by wind and water can eat into the firm tangle of tough root, leaf, and rhizome that may honeycomb the entire mass of a dune.

Dune building is the work not of waves but of wind. Loose grains of coarse, dry sand are easily moved by a stiff breeze, and when a beach becomes so high or so broad that part of it is more or less permanently dry, any little projection that drags at the wind and intercepts the rolling sand can touch off the formation of a dune. A stranded log or even a clump of beach grass may serve the purpose.

One can watch the process in miniature in a spot where the wind is raising ripple marks on a sandy surface. Sand grains go rolling up the gentler windward slope of the ripple, then topple down the little lee slope and come to rest at a steep angle. The same processes operate on a larger scale to form dunes that may rise thirty or forty feet high if there is an abundance of sand available for building material. As the wind shifts and eddies change their positions, the direction of sand movement changes, too; and the forms of dunes may be anything from a single long, low ridge to a chaotic tumble of steep, irregular hills.

If dunes are built up faster than they can be anchored by the growth of vegetation, they may migrate inland, wave after wave, before the wind, engulfing and killing even large trees, then exhuming the skeletons as the dune moves on. Not very many places on the New England coast have enough loose sand to make large dunes; but the Ipswich region near Cape Ann has a large expanse of real duneland,* as do several places on Cape Cod.

Even a modest expanse of sand above the reach of the waves may bear a characteristic kind of vegetation. Among the clumps of beach grass grow the trailing beach pea, fleshy-leaved seaside goldenrod, and evening primroses and blazing stars. In spots sheltered from the wind, very often in hollows on the lee sides of dunes, shrubs may grow into a dense thicket. Here you will find the beach plum that shows as a tuft of crooked black sticks in winter and a cloud of white blossoms in spring. In late summer it produces in their millions the fruits that are made into a tart jelly of which the chief claim to distinction is probably not so much an exceptional flavor as its association with the haunts where the beach plum grows. The sandy thicket may also include bayberry, sumac, wild roses, and alas! poison ivy. Where conditions moderate enough for trees to grow, the shrubbery gives over to woods of oak or pine.

* These dunes and nearby marshes have been described in loving detail by Dr. Charles Townsend in a book called "Sand Dunes and Salt Marshes," published in Boston in 1913.

In the estuaries of our largest rivers and in countless other places sheltered behind a barrier beach or up some tidal inlet, great expanses of marsh have developed. Where a river pours large quantities of fresh water into the sea, there may be a gradual transition from true saltwater marsh at the river mouth through brackish water to fresh water upstream that rises and falls as it is alternately held back and let out by the rise and fall of the tide. Vegetation in the freshwater parts of such an estuarine swamp is much like that of any freshwater swamp.

More numerous and generally easier of access are the true tidal salt marshes. The seaward edge of a marsh is normally a soft, dark muddy flat exposed at low tide and giving off vapors that in any other context would be considered obnoxious. The mud flats are too dry for sea plants and too wet for land plants; and when it is not flooded, the mud lies bare and sodden. About at the half-tide line a lush vegetation begins abruptly; and from here back to the upland woods it forms a series of clearly marked zones that are the result of differences in saltiness and tide level.

Facing the open water and bordering the edges of creeks and ditches there is often a little bluff that rises a foot or so above the edge of the mud flat. On the slope of this bluff, between half-tide and normal high-tide level, grows a pure stand of tall, coarse salt thatch or saltwater grass *(Spartina alterniflora)*. Where the slope is gentler, the tall grass may form a band a yard or two wide paralleling the water. On sharp slopes one can often see that the wet, peaty soil in this zone is being eroded along the channel edges wherever a rapid current touches it.

Above the band of saltwater grass there is an abrupt transition to smooth, almost level meadow that is barely awash during the highest stage of normal tides and thoroughly drained between tides. This gentle slope is covered with a thick sward of grassy vegetation that is marked off from the adjoining zones by its shorter length and finer texture. The lower part of the sward is covered with salt meadow grass *(Spartina patens)*. This is the kind that makes the choicest marsh hay. It can be recognized by its bright, yellowish color and soft texture and its tendency to become matted into swathes and cowlicks. The upper meadow is covered with the somewhat more brownish and darker blackgrass *(Juncus gerardi)*, a kind of slender rush. Over the meadow may be scattered such little flowering plants as sea lavender, asters, gerardia and the seaside goldenrod. Some marshes may grow up thickly to the shrubby marsh-elder.

At low tide the meadow is reasonably dry, and where ditches are not too wide for safe jumping, one can walk far out across the springy turf over a peaty mass that may be a few inches to many feet thick, composed of the remains of dead grasses and rushes mixed with silty mud.

Scattered here and there over the marsh are shallow depressions, or "pannes," of various shapes and sizes, many of them waterfilled, some in different stages of drying. A panne may be bare of plants, with its muddy bottom drying down to salt crystals; or it may hold a central puddle surrounded with much-stunted saltwater grass that can be recognized by its broad if diminutive leaves. Glasswort or samphire may occupy a dryish panne by itself. Other pannes make bright patches on the meadow when their sea lavender or gerardia or asters are in bloom.

Where the land surface rises at the edge of the marsh, whether against an upland hillside or around a boulder island, the meadow ends abruptly in a narrow strip of tall switch grass *(Panicum virgatum)*. Whether or not you consider this strip to be part of the marsh is a matter of definition. It has the peaty soil of a marsh, but the plants growing with the switch grass are an odd mixture of tidal marsh, freshwater marsh and even woodland species—marsh-elder and seaside goldenrod perhaps growing with red maples and turk's cap lilies. On soil that rises above the influence of tide water a narrow shrub border faces down a woodland of whatever kind is common in the region.

Man has been interfering with natural events in the marshes for a long time. Probably the effect of Indian activity was transient and slight, mostly limited to fire in the season when grasses are dry. But

European settlers found the salt meadows a great boon for their grazing cattle and for hay for winter feeding and bedding. Stone walls running out over the marshes date back to the early days when well-marked property lines were important on the limited meadow area. Even today salt hay is prized as a garden mulch, since it does not mat in the winter weather and is free of weed seeds.

In recent years large areas of marsh have been crisscrossed with straight, narrow mosquito-control ditches. These are intended to improve drainage and eliminate breeding pools, and indeed their first effect is to drain the marsh and lower the general water level. However, natural processes are set in motion that soon change the situation.

In some cases the large saltwater grass may invade and grow so vigorously that it soon fills the ditch with peat. On the other hand, if tide water flows in and out very rapidly, scouring by the current may considerably enlarge the ditch. An enlarged ditch behaves rather like a natural creek, where water rises and falls with the tide, making the channel a small version of an estuary.

As silt-laden water flows up a channel on the flood tide, it rises higher and higher until it overflows the banks and spreads out shallowly over the marsh surface. This checks the current sharply, and just as when a river overflows its flood plain, a new layer of silt is deposited, with the thickest part of it right at the channel edges. As a result, both tidal creek and enlarged drainage ditch become bordered with natural levees. Along the ditch an additional levee effect is produced by the blocks of turf dug out and piled alongside in a little ridge. Whatever its origin, a levee interferes with drainage of the marsh surface lying between the ditches or creeks, and new pools are eventually formed. It is clear that ditching, whatever may be its immediate effect on mosquitoes, is far from a once-and-for-all proposition.

One might ask why there are so many marshes along the Atlantic seaboard. Like the marshes themselves, the answers are somewhat different for New England and for other parts of the coast. To the north, in the Bay of Fundy, silting is much more rapid and more important in the process of marsh development. South of New Jersey the geological history of the shoreline is quite different from that of the New England coast. The so-called "New England type" of tidal marsh owes its existence to the relatively recent subsidence of land—or rise of sea level, whichever way you choose to look at it.

In this region a marsh begins when a sand bar forms across the mouth of a sinking bay, enclosing a lagoon of quiet water where fine silt and mud settle to the bottom. Some of the silt is brought in by streams flowing down from adjacent uplands; some may come in on the tide, gathered up from a source that may be miles away along the shore. When a deposit of silt grows thick and high enough so that it is alternately flooded and drained by the tide, saltwater grass begins to move in.

The appearance of grass has two effects. The dense stand of leaves reaching up into the tide water serves as a trap for incoming silt and sand; and in the waterlogged deposit of mud that forms, dead grass decays slowly, so that its remains accumulate over the years to make a fibrous sort of peat. As a result of all this, the developing marsh surface is built higher and higher. When it reaches the level of high tide, salt meadow plants come in, and a typical marsh is under way.

As the peaty soil builds up layer on layer, it keeps a record of all the various things that fall on the surface of the marsh. First there are the plants themselves. Even when their softer parts have rotted away, enough remains of the woody, fibrous tissues so that an experienced person can recognize at least the general type of plant from which they came. There may be layers of sand washed up by exceptionally high storm tides, or ash from a forest fire in the vicinity. There are bones or shells of animals that have died on the marsh. Then there is each year's deposit of pollen grains, just as in a bog or lake bottom.

Of course there are always people curious to see what they can learn about the past by whatever means, and marshes, like bogs, have been probed for their secrets. If one goes far enough down through the marsh peat, he eventually works through the plant remains and strikes purely mineral soil that quite naturally resembles that of the adjacent dry land, in New England commonly glacial deposits. Above this there may be as much as fifteen or twenty feet of silty peat, all formed since the departure of the ice sheet. As one might anticipate, the top layers of peat consist of remains of salt meadow grass, the kind used for hay, and below that are remains of the coarser water's-edge saltwater grass.

The surprising thing is that below the saltmarsh peat one ordinarily comes upon freshwater peat and even stumps of trees still rooted in the mineral soil—not at all what one would expect to find in marine deposits. In a number of places along the coast, especially in Connecticut and on Cape Cod, rooted tree stumps have been found buried beneath several yards of marsh peat and standing as much as sixteen feet below present high-tide levels.

Now, in interpreting the past history of a marsh, it is not possible to tell just how many years are represented by a given thickness of peat. Just the weight of the top layers is enough to compress the lower ones to an unpredictable degree. In one place, for instance, it was found that the weight of a sand bar that had moved in part-way over a marsh had compressed a fourteen-foot deposit of peat into a tough, leathery layer only four feet thick. There are other complications in interpreting specific deposits, and there is strong geological evidence that there has been no appreciable change in sea level for three or four thousand years. Still, the conclusion seems unavoidable that sea level has risen since the trees whose remains we find were part of a living forest of no great number of centuries ago.

It appears that the sea is still rising. Careful studies of present tide levels all along the Atlantic seaboard from Maine to Florida lead to this conclusion. Figures recorded in the years since accurate measurements have been made show that from 1895 to 1930 there was an extremely slow but steady rise in sea level, and that the rate of rise has increased since then to about two-hundredths of a foot per year. Assuming that this rate of change continues, two feet in a century will not much concern even our great-grandchildren. But if our seaboard cities last long enough they may eventually have to adjust their waterfront activities to a rising ocean.

Another change that can be expected in the course of time is the progressive filling in and disappearance of coastal marshes. Silting and peat formation act constantly to raise the level of a marsh even as the land subsides. At the same time waves and wind continue to pile up bars and dunes and then drive them inland over marsh and lagoon. When the soft peaty soil of a marsh is no longer shielded from the ocean by beach or bar it is easily eroded by the waves. As a result the beach line moves farther and farther in until it reaches the edge of solid, rising ground. Then the surf eats into the land to form a long, straight line of sea cliffs such as the ones that face the Atlantic along the outer end of Cape Cod.

Through the centuries this sunken, irregular seacoast has provided a living of one sort or another for the men who have lived there. Today's traveler is struck with the great numbers of large, handsome houses in town after town along the coast all the way from Maine to Connecticut. The best of the houses are well over a century old, and most of the towns look as though nothing much had happened there in that length of time. Where in the world, one wonders, did the money come from to build them all?

Of course the answer is that it came from all over the world—from the China trade and the Indies and the whale oil industry. The hilly, glacier-scoured hinterland offered only limited possibilities for agriculture, and milltowns were generally located farther inland where rivers run fast and steep; but forests everywhere were full of oak and pine for building ships, and the jagged coast held countless sheltered deepwater inlets for shipyards, wharves and warehouses. So the people turned early to the sea, and for a long time it gave them a vigorous prosperity.

Then within the last century steam and steel and centralization brought an end to the bustling maritime life of the hundreds of little coves and inlets. Today the fine houses remain to ornament the landscape; the sea with its bays and its offshore banks provides a modest living for fishermen and lobstermen, and the larger towns still have moderately active lives as lesser seaports. But to thousands of Americans the New England coast means chiefly happy summer days spent sunning on the beaches or "messing about in boats."

8

The Oak, Pine, Sand Country

WHEN CAPTAIN JOHN SMITH published "A Description of New England" in 1616, he said of the forest there: "Oke is the chief wood; of which there is great difference in regard of soyle where it groweth." Cape Cod he dismissed as "onely a headland of high hils of sand ouergrowne with shrubbie pines, hurts,* and such trash." Today's traveler finds the situation little changed in southern New England.

The forest there bears more resemblance to that of Long Island and the south than it does to the New England north woods. The birch, beech, sugar maple, and white pine trees so characteristic of the north are few and scattered. Instead the most common and conspicuous and universally present species are oaks and hickories, and in its day the chestnut. The poorest sandy soils usually grow up to pitch pine, but the texture of most of the wooded landscape is the dark, glossy green of broad oak leaves.

Much of the present oak forest is rather dry and open. This is quite clearly related to fire, but ecologists are not entirely agreed as to whether fire is the cause or the result of the open nature of the woods. One line of argument holds that the more northerly plants that need cooler and moister conditions have been kept in abeyance for centuries by fires set first by the Indians and later by Europeans.

* Hurts: whorts, whortleberries, blueberries.

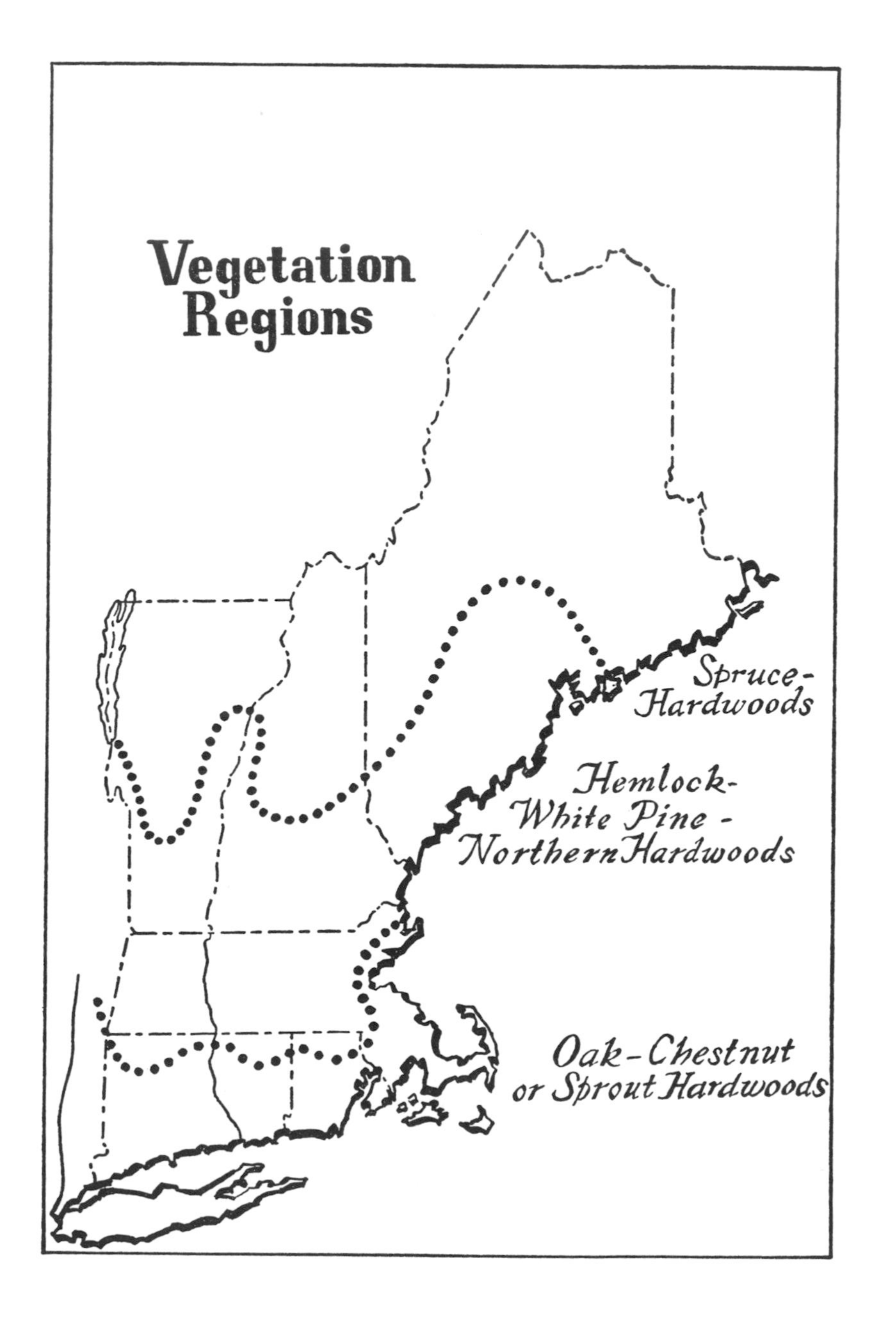

Vegetation Regions
Spruce-Hardwoods
Hemlock-White Pine-Northern Hardwoods
Oak-Chestnut or Sprout Hardwoods

Another point of view considers it preposterous that fires could be widespread and frequent enough to control the vegetation over hundreds of square miles for hundreds of years.

There is evidence though, that fires need not be very frequent to maintain their influence on the forest. In the year 1845 a severe fire burned over one shoulder of Great Mountain, in the northwestern corner of Connecticut. Regrowth was fast enough so that this "Mansfield burn" tract could be cut for charcoal in 1863 and then again before the turn of the century. Ninety-nine years after the great fire, and with two clear-cuttings in the meantime, foresters found that the outline of the old burn was still clearly defined both in aerial photographs and from observations made on the ground. One of the most interesting distinctions was that hemlock is common on the surrounding un-burned area but stops short at the boundary of that long-ago fire.

From the present state of our knowledge, it is not clear why, fires or no fires, the northwoods type of forest does not cover all of New England, at least after fires have been stopped. There is some indication that when the woods are allowed to go their own way undisturbed for a long enough time, the northern trees do in fact become more abundant wherever the soil is moist enough and there is shelter from drying winds. Shady ravines, north-facing slopes, and land lying just above swamps may have a much wider variety of trees than the drier oak forest on higher land; and in such places sugar maples, beech, hemlock, and black birch grow along with the oaks and hickories. The soft, dark hemlocks especially increase to the point of abundance, even in southern Connecticut and Rhode Island, wherever there is shelter from the hazards of fire, wind, and hot summer sun.

The most credible explanation for the presence of the oak forest in New England seems to be that it moved northward from the Appalachians several thousand years ago, shortly after the retreat of the glacier, during a period when we know that climates in this part of the world were generally warmer and drier than they have been in the past five hundred years or so. In this view the oak forest in our region is a relic of conditions that no longer exist. It is understandable that changes in vegetation, especially trees that may live for centuries from the time they start as sprouting seeds, should lag behind changes in climate; and one might speculate on what will happen if the warming trend of the past fifty years goes on for several centuries.

No two people draw the northern boundary of the oak forest in just the same place; but all agree that the line is irregular and indistinct. It apparently lies a little to the north of the boundary between old-field red cedar and old-field white pine. The traveler can follow the example of a forestry professor who amused himself during a year of commuting between Boston and New Haven by driving a different route on each trip and marking the pine-cedar boundary on a map. When the border region is mapped in any detail, the boundary line is found to be very irregular, and the reasons for its ins and outs are not always readily apparent. Temperature, moisture, topography, and soil all go into the formula, and sometimes very slight differences must be enough to tip the delicate balance one way or the other.

Hardly any remnants survive of New England's oak forest in its primeval condition, and everything considered, this is no wonder. It covered the parts of the country that have been settled longest, being most accessible by water. Nearly all of the area has at one time or another been cleared for agriculture, and much of it has since been abandoned to grow up again to a young, scrubby woods.

The better places, where the soil is fertile and neither too wet nor too dry, originally supported a varied mixture of oak, maple, and chestnut, with a scattering of others such as basswood, sweet birch, ash, and the tall, straight tulip poplar. Early settlers soon learned that this combination of trees grew on the kind of soil that produced the best crops; so most of it fell early to the ax, and much of the land where it grew is still under cultivation. Where the soil is somewhat poorer and drier, the commonest trees are black and white oaks and hickory, which early records call "walnut." The driest ridges are left to the red and chestnut oaks and the few sprouts that persist of the chestnut itself. Soil that is too thin and too dry to support a real forest and bears only scattered trees is the original home of the light-demanding red cedar, or savin.

Below the over-arching canopy of tall trees there is commonly a lower layer, formed of the crowns of smaller trees and shrubs, some of them youngsters of the forest giants, most of them other kinds that never grow very large. A very common part of the understory is that paragon of small trees, the dogwood, that each state from Connecticut to Georgia seems to feel belongs to itself above all others. Along with the dogwood grow sassafras and hornbeam, laurel, blueberries and hazel. Then down on the floor of the woods there is a carpet of little herbaceous plants—wintergreen, partridgeberry, Canada mayflower, and trailing arbutus among the most common.

The presence of many oaks, with their tough, leathery leaves, may strongly influence the soil of the forest where they grow. In poor places where almost every tree is an oak, the earth is covered all through the year with a matted layer of leaf litter that is sharply separated from the mineral soil beneath it. Oak leaves contain so much tannin—the same chemical stuff that is used to harden and preserve leather—that they are resistant to the action of decay microbes, and earthworms have no taste for them. So the leaves accumulate; and in extreme cases they may develop into a thick, almost impenetrably tough, fibrous mat where very little else can grow save blueberries.

Where many other trees grow along with the oaks, the soil is in much better condition. Unlike those of oaks, the soft leaves of trees such as maple, ash, and basswood decay readily, and earthworms relish them for food. As a result, each year's crop of autumn leaves disappears into the soil in a few months, and by midsummer there is practically no loose organic litter left on the top. Instead, all the leaves have disintegrated into a fresh lot of minute humus particles that filter down into the earth to make a dark, crumbly topsoil. Such a soil is loose and porous and at the same time holds moisture so well that it does not readily dry out. It is loaded with mineral salts that were drawn up through the tree roots from deep in the soil and built into leaf and stem tissues. The leaves that fall every autumn make a rich top dressing that soon decays and works down into the earth, steadily improving the texture and fertility of the upper soil in a sort of self-propelling cycle.

A fertile "mull" soil of this kind makes excellent farmland, and little of it is left in the wild except where the land is too steep or too rocky for cultivation. In the springtime the ravines and hillsides where it survives are carpeted with a wide array of little flowering plants. Almost anything that can grow in shade and can hold its own against heavy competition can grow in such places, and a handful of our spring flowers can grow nowhere else. Jack-in-the-pulpit, anemone, Dutchman's breeches, toothwort, bloodroot, wild ginger—all these are sure clues to a good mull soil. None of them can grow in the heavily-matted oak woods, and the spring carpet there consists rather of starflower, goldthread, partridgeberry, and great masses of Canada mayflower, or "wild-lily-of-the-valley."

To see the May flowers that April showers bring, one must go to the woods while tree buds are just starting to swell. Here the bright sunshine passing freely through the still leafless branches onto the forest floor rapidly warms the earth in early spring; and while deeper roots and loftier boughs are still surrounded by wintry chill, small plants that live their entire lives within the few inches just above and below the soil surface are roused by the first returning warmth.

This special world of the broad-leaved temperate forest floor harbors a great horde of small, spring-blooming plants that have adjusted their rhythm of life neatly to it. From the first hepaticas to the last pink ladyslippers, they start early and finish most of their year's business before the trees overhead have any more than opened shop for the season. Leaves and blossoms were long since formed in miniature, and the spring's rapid burst of growth is only the carrying to completion of processes begun the summer before. Flowers open promptly with color and fragrance that are welcome signals to early-prowling pollinating insects. New leaves soon follow, unfolding their green manufactories that fashion from simple elements of soil, air, and water the infinite complexities of root, shoot, and blossom.

By early June, when the tree leaves expanding overhead begin to shut out most of the sunlight, flowers have gone to seed, reserves have been stowed away in root or rhizome ready for the first rush of next spring, and the germs of next year's leaf and flower are already formed, tightly curled in buds that will lie dormant through the autumn and winter. Some of the small woodland plants then go quietly to sleep and disappear by early summer. Sometimes a few leaves persist until autumn, inconspicuous in the deep shade of the high forest canopy.

This broad-leaved temperate forest is almost unique among forests of the world in its dramatic change with the turning year. Each season has its own charm: the flowers of spring, the green richness of summer, and the flare of autumn color. But it is in winter when most of the trees are bare that this long-settled land reveals clues to the story of its past. Most conspicuous and most eloquent are the stone walls that outline roads and fields and woodlands everywhere in New England. Proper New England walls are dry-built, not rigid with mortar, but kept in place only by skillful arrangement and balance of their rocky elements. Sometimes the stones have been cut and shaped so that only the smallest chinks interrupt the flat surfaces of top and sides. But true country walls are made of rough stones just as they were carted from the field in an ox-drawn stone boat and piled into straightforwardly functional fences, full of sheltering crannies for mice and chipmunks.

The miles upon miles of wall that the winter traveler sees from the road are the accumulation of two, even three centuries of labor. From earliest colonial days the building and upkeep of fences was one of man's most important private and civic duties; and considering the local situation, fence usually meant stone wall. Public records of the colony of Connecticut show that the General Court strove mightily with the problem of proper fencing. Repeatedly they handed down regulations intended to compel farmers to maintain fences that would be adequate to keep cattle on the property of their owner and more especially out of the neighbor's cornfield.

In 1643 after a number of individual judgments, the Court ordered that each town should forthwith "chuse fro among theselves seaven able and discreet men" to ponder and make recommendations for improving the common lands. "And whereas also, much damage hath risen not only fro the unrulynes of some kynd of Cattell but also fro the weakness & insufficiency of many fences, whereby much variance and difference hath followed, which if not prevented for the future may be very prejudiciall to the publique peace; It is therefore likewise Ordered, that the said 7 men soe chosen, or at lest 5 of the, shall sett downe what fences are to be made in any Comon growuds, and after they are made to cause the same to be vewed, and to sett such fynes as they iudge meet uppon any as shall neglect or not duly attend their Order therein. And when fences are made and judged sufficient by the, whatsoever damage is done by hoggs or any other cattle shall be paid by the owners of the said cattle, without any gaynesaying or reliefe by Repleivy or otherwise."

Even this firm dictum did not settle the matter, and later Courts had to order the fence viewers time after time to tend more conscientiously to their duties. More than a century and a half later the office of Fence Viewer was still important enough to be incorporated into the governmental machinery of new states such as Ohio.

For generations wall building went on as fields were cleared of rocks and trees, until most of New England became laced over with a fine-meshed network of stone. Though the walls often run with no discernible meaning through the woods today, they mark off what once were open fields and lanes, or show the course of the local road before it was leveled and straightened to suit the demands of traffic moving so much faster than the horse or the ox. Many a roadside picnic area makes use of a nook left by a rounded-off curve or a relocated bridge, and one of its charms may be the wall that still separates private field from public way.

In this land the trees, too, record details of local history. The ordered row of wide-spreading maples edging the woods by the roadside once graced the front of a house that may have fallen before that horror, fire in the country. That ancient oak with its low, heavy boughs forming a crown as wide as it is high clearly lived its formative years in the open, without the jostle of slim young things that now crowd around its knees. Long ago it was left in an open pasture to shelter the cattle from sun and storm, and it had grown to stately size when the farmer gave up the struggle and the brush crept in. The tall red cedars now surrounded by younger trees certainly started in an open, grassy place a long time ago when there were no fast-growing broad-leaved trees nearby to overtop them and shade them to death.

Many patches of woods have no ancient giants among the smaller trees; but look at the way the trunks grow from the ground. Nearly all of them stand in bouquetlike clusters. Perhaps there is still a remnant of the stump in the center of each group from which the present trees sprouted when the woodlot was clear-cut, something between fifteen and forty years ago. The parent woods, too, may have originated as coppice or sproutwood that grew up after an earlier woodland was cut for charcoal and posts and cordwood; and the present trees may be as much as the fourth or fifth generation of sprouts from the original seedling trees.

Fires and grazing, too, leave their marks on the land. An abundance of young red cedar means that a poor sort of pasture is being invaded by trees. Cattle eat the seedlings of broad-leaved trees as fast as they appear, but leave the prickly cedars strictly alone. Fire, on the other hand, kills red cedar but encourages the increase of the fast-growing black cherry, which only sprouts more vigorously from every root and stump when its above-ground parts are destroyed. Aspen and the little gray old-field birch seed in on bare soil when an old cultivated field is abandoned or when a hot fire sweeps through a dry woodland, burning off the protective humus from the ground.

Even without human intervention, any forest is disturbed in one way or another over a long period of years. Since 1938 any list of natural disturbances in New England includes hurricanes. That historic storm provided many opportunities to learn just what effect a violent wind has on a well-grown forest. One instance was that of an old stand of oaks and hickories that grew on a hill near the shore at Stonington, Connecticut. The area had not been disturbed for a long time, although it was probably not a truly virgin forest. Fortunately it had been studied carefully in 1913; and after the tremendous storm twenty-five years later it was examined again in detail, before the windblown trees had all been cleared away and while there were still many stumps available for counts and measurements of the annual growth rings in the wood.

Study of the tree rings showed that large trees on the seaward side of the woods bore a clear record of another destructive hurricane that had struck there in 1815. Working backward from 1938, tree after tree showed good annual growth for a hundred and twenty-three years. The growth rings for 1816 were especially wide. At that point the rings abruptly became very narrow, indicating poor growing conditions, for a dozen or so years right back to the seedling infancy of the tree.

The story implicit in these growth rings is that the trees started life on the floor of a well-established forest. For years their growth was greatly suppressed by crowding and by the shade of the old trees above them, so that each year they formed only a small amount of new wood. When the large trees finally went down in the wind, all the small ones were suddenly and simultaneously released from suppression and made a great spurt of growth, laying down a thick band of new wood every year until they in their turn were blown down.

On the landward side the forest appeared to have suffered less from the wind, both in 1815 and in 1938. Instead of the even-aged stand that developed where the old forest had been badly damaged, trees of all ages were present on this side of the woods. The oldest trees were of the same kinds as the younger ones, mostly black and white oaks, hickories and red maples, showing that the composition of the forest was not changed even by the devastating force of a hurricane. It also suggests that the forest was reproducing itself in kind, and that successional changes had reached the stage where climate and soil determined the kind of plants that would grow—in other words, that this was the climax vegetation for the region.

Other less dramatic but more long lasting natural catastrophes have struck the forests of southern New England. In the course of thirty years a microscopic but virulent fungus transformed the chestnut from an abundant and valuable timber tree to a minor part of the underbrush. The chestnut blight disease was first noticed in 1904 on a few trees in New York's Bronx Zoo. By the time it was recognized as a serious pest it had spread into all the neighboring states, and before it was well enough understood to allow control measures to be devised, the chestnut had become virtually extinct.

One might think that the sudden disappearance of a large tree that had been abundant throughout half the country would make a drastic change in the appearance of the forests. Actually, the change has had its greatest impact on users of the chestnut—men who used its wood for timber, and men and animals that used its nuts for food. Chestnut trees were very common, but they grew rather scattered in the old forests. As they died off, they left relatively minor gaps in the forest canopy, and their living space was soon taken over by the enlarging crowns of nearby existing trees, mostly oaks. Consequently their passing has made no dramatic change in the general aspect of the forest.

More drastic than natural catastrophes has been the effect of cutting on the nature of the forests. Land entirely cleared for cultivation or pasture and then abandoned undergoes a long series of changes that have already been described. A forest that is clear-cut but where no stump removal or burning is attempted also soon grows up to woods again. In this case the new generation of trees will consist of those that are already present as seedlings and those that sprout vigorously from the old stumps.

If the forest is cut repeatedly its composition gradually changes. Cedar, hemlock, and most pines do not sprout at all, and one cutting eliminates them entirely. Chestnut has been essentially destroyed by the blight. The more desirable oaks sprout, but they grow relatively slowly. The long-term result of all this has been a great increase in the proportion of such fast growers as red oak, black cherry and red maple. In low-lying wet places sproutwoods may be reduced over the years to a nearly pure stand of rank-growing red maple. This tree is vigorous enough, but practically every bole becomes rotten at the heart, and it is good for nothing but an indifferent sort of fuel. On drier ridges repeated cutting in time converts a coppice to pure oak woods. Before its extinction chestnut could keep up with the oaks; but now it is coarse and fast-growing red oak that forges ahead and overtops the hickories and maples to spread out above as a "wolf tree."

There are a few forests that have never been clear-cut; but even these have been subjected to selective logging. This, too, is an effective way to remove the good trees and leave the trash to accumulate. Foresters say that the better kinds of trees are still present in large enough numbers to restock the woodlands without recourse to either artificial seeding or planting. It is only necessary to handle the woods in such a way as to favor the more useful species.

The greatest hazard to forest improvement seems to be premature cutting. In southern New England woodlands are owned in small parcels by small operators who can scarcely be expected to think in terms of fifty year rotations, and most often the wood is cut for whatever it is worth whenever the owner needs a little loose cash. The scanty yield of wood that can be extracted goes mostly for fuel, although some is used for posts, charcoal, and railroad ties.

A traveler in this part of the country has an impression of countless tracts of rather young, scraggly woods that amount to very little except for their large total acreage. Statistics from a postwar survey of the forest situation confirm this impression. For example, of the forested area of Rhode Island in 1944:

30% was too young to yield even firewood;
33% could produce a little fuel from the larger trees;
20% could yield fuel and a little low-grade saw timber;
1% bore stands of saw timber, but this was so scattered that its existence was more theoretical than actual.

In Connecticut 90% of the forest was less than sixty years old, half of it less than twenty. The problem has been described in an acorn-shell as "too little wood on too many trees."

Here and there in the larger oak forest are smaller tracts covered with the three-needled pitch pine, with or without a mixture of oak. The pine woods grow mostly on flat expanses of sterile, sandy soil in old postglacial lake beds and in the hummocky terminal moraine country. Farther north white pine grows on wet, sandy "pine flats" or dry, gravelly terraces; but we are concerned here with the pitch pine woods that are scattered through central Massachusetts and Connecticut and cover large areas in the vicinity of Plymouth and on Cape Cod.

Much of the Cape and also Nantucket were originally covered with oak forest that was interspersed here and there with pine. This disappeared long ago in the face of cutting and grazing. The Elizabeth Islands, when Captain Gosnold and his men first saw them in 1602, bore grassy meadows full of berries and set about with "high timbered oaks." The same discoverers named Martha's Vineyard for the tangle of vines that then covered a large part of it.

From the earliest days of colonization, pitch pine forests have been exploited both for their wood and for its resin content. Trees were "boxed" or tapped for turpentine and other forms of naval stores. Knotty, resinous faggots were formerly used as "candle-wood." Large inroads have been made for fuel for domestic use; and in the early nineteenth century huge quantities were used to fire the boilers of wood-burning locomotives.

A pitch pine forest is dry, rather open, and highly combustible. The pines themselves are not seriously damaged by light fires. In fact, this species is one of the rare conifers that will sprout from a stump when its top is damaged or destroyed. Repeated burning eliminates the other kinds of trees that grow among them, however, leaving the pines in possession. If the pines are removed, or if fires are kept out, black oak tends to take over in the long run. Open parts of the wood may have a shrubby underlayer of sweetfern, bear oak, New Jersey tea, and other plants that can survive in dry, sterile soil.

There are apparently more trees on Cape Cod today than when Thoreau went there in 1849 and 1855 to make his acquaintance with the ocean. But the Cape, for all its dense summer population, still harbors a wild sort of vegetation that is not easily found anywhere else in this thickly settled end of the world. For this remnant of wilderness we have largely to thank the inability of the soil to support any more utilitarian plants, except for the cranberries in their cultivated bogs. Where neither oak nor pine can grow, bayberry, sweetfern, beach plum, and the undemanding blueberries fill in. Creeping on the ground there are neat, compact, evergreen bearberries; delicate, trailing cranberry plants that look too fragile to bear so robust a fruit; and shrubby little woolly Hudsonia, sometimes called "poverty grass" because it is the last to give up the struggle against adversity.

On high cliffs such as those that stretch for miles along the "forearm" of the Cape, where the soil is quite adequate to support an oak forest, strong winds blowing from the ocean are in control. Where the conformation of the headlands diverts the wind up a little above the land surface, or in sheltered hollows, there is a true oak forest, but one kept eternally pruned low by the wind's blast. Any tree that sends a hopeful shoot upward during the mildness of summer is sure to be nipped back by driving sand or sleet during the winter. Only the twigs that lie low can survive. So the woodland becomes a densely matted tangle of tough little trees, a few inches tall where the wind hugs the ground, reaching up perhaps several feet in hollows where the land drops a little below the ceiling of wind.

Old Captain John Smith would find incredible changes in the

New England that he knew. And yet, from the windswept headlands of Nauset that face the open Atlantic to the sheltered hillsides of the Housatonic, it is still true that "Oke is the chief wood;" and his observation still holds that "there is great difference in regard of the soyle where it groweth."

9

The North Woods

DRIVING NORTHWARD from the coast on any upland road in Connecticut a traveler gradually becomes aware that a northern look is coming over the land as the slim red cedars of the scrubbier pastures to the south give way to dark, spreading white pines. A little farther along, a glance at a well-grown woodland confirms the impression of change. Oaks and hickories are now far less common. Instead, sugar maple, birch, and beech are becoming abundant. Birches are not always the meager little gray ones, but may often be tall, handsome specimens of yellow birch or the beloved, truly white-barked paper birch.

This is the beginning of the north woods, the forest that reaches north across upper New England into the Saint Lawrence Valley and Nova Scotia, westward across the upper Great Lakes, and even sends a narrow arm down along the higher slopes of the Appalachians as far as the Great Smoky Mountains. Different people have seen different aspects of this forest and named it accordingly. Probably the name that gives the clearest picture of it is the longish descriptive one of "Hemlock-White Pine-Northern Hardwood."

The terms "hardwood" and "softwood" are used in a rather misleading way by lumbermen, and their usage is followed by foresters. In this special sense, hardwoods are the trees with broad leaves that are shed every autumn and renewed every spring—in other words, deciduous trees. Softwoods are chiefly the needle-leaved, cone-bearing evergreens. The terms were probably used originally for such trees as the hard sugar maple and birch and the soft white pine and hemlock, but there are many "hardwoods" whose wood is actually soft, and vice versa.

To New Englanders, a forest of hardwoods, with a few needle-leaved evergreens, seems the normal kind of vegetation that should cover the earth. Actually, such vegetation is limited to a rather small part of the world's land surface; all the rest bears either coniferous or broad-leaved evergreen forest, or else grassland or desert.

Of all the charms of the hardwood forest with its changing seasons the most dazzling is the bright display of colors that it makes each autumn as the leaves begin to fall. The autumn spectacle is very nearly a monopoly of eastern North America. The forests in parts of Europe and eastern Asia are potentially able to put on a similar show; but Europe has relatively few kinds of indigenous trees, and even those do not reach their full brilliance in the damp and cloudy oceanic climate. The temperate forests of Asia are said to rival North America, although not very many westerners have been in the mountains of China in October to see them.

In our country the various kinds of hardwood forest extend in a great triangle from the southernmost Appalachians to Minnesota to Maine and across the adjoining parts of Canada. The autumn performance, though, seems to reach a pinnacle in New England. Here the wide range of colors of the many kinds of native trees, including a proportion of evergreens to provide dark accents, combines with a special clarity and crispness of the atmosphere to intensify the effect, and the bright tapestry is spread over hills and valleys that roll it up into wide, sweeping vistas.

The display of color that delights even the stolid human eye serves no known purpose for the trees. All summer long the leaves are green with their abundant chlorophyll, itself a highly utilitarian substance. Then as the living tissues age and become less active, and as cooler weather slows all vital processes, used-up chlorophyll is no longer replenished. This unmasks the more lasting yellow pigments that have been present all summer, but concealed by an overlay of darker green.

When days are still warm and nights begin to cool, but well before the first frost, the reds begin to appear. Unlike the yellows, the redness is something added as the autumn comes on. Just what this means to the trees is not fully understood, but it seems to be a method of disposing in a harmless way of excess sugar that accumulates in the moribund leaves. The bright reds are lacking from the autumn woods of other parts of the world. Even in New England not every kind of plant has the ability to turn red; but it is the very scattering of talent that makes the show so varied.

Another contributor to the autumn pageant is the brown substance, tannin. This is found in many plant tissues, although only the bark of a few kinds of trees contains enough to be useful for curing leather for shoes and saddles. Tannins are the most durable of the colored substances, and when green and red and yellow have finally faded away, the browns remain, even when leaves and twigs have fallen in crumbles and gone to form part of the soil.

From this handful of colors an endless variety of tints and hues is assembled in the leaves each autumn. Many shades are so characteristic that with a little practice one can recognize a tree or shrub from as far away as its true color can be discerned. The prize for clarion scarlet surely goes to the sugar maple. Other plants turn red, but none quite equals the inner light of a sugar maple in full gold and scarlet glory. Yet I know a grove of these—now, alas! almost defunct—that every year shows no trace of red. Even on a gray, rainy day walking among the trees there is like walking through solid golden sunshine.

Almost a match for sugar maple are sassafras and the treacherous poison ivy that romps with so little restraint over walls and up trees and through grass. Sumacs flare in the roadside brush, and swamps are touched with the glow of red maple and highbush blueberry. Ash and dogwood have their distinctive purples. There is a bright, golden brown that always turns out to be a beech, the only brown that seems like a truly bright color. Nothing but hickory can produce a certain special shade of pure antique gold.

Even while some of the leaves are still brightening, others have already begun to drop. Preparation for autumn's leaf fall begins far back in early summer, almost as soon as the leaves are full grown. Just where a leaf attaches to its supporting twig a layer of tissue develops that forms a slight discontinuity, a potential zone of weakness. All through the summer the leaf manufactures hormones, regulative substances, that hold in abeyance the development of the region of weakness. But in the cool of autumn, hormone production tapers off, and the tissues gradually separate from each other until the leaf is held only by the strength of the veins that run from leaf into twig. It takes only one night of wind and rain to break these fragile connections by the millions; next day the splendor of the woods is gone.

When the leaves have fallen, the land takes on an appearance of winter sleep. Actually, the year's "rest" begins long before leaf fall. Though vegetation remains green into September and even October, new growth of trees takes place almost entirely in spring. As early as midsummer the next year's buds are all formed and the year's layer of wood is laid down in root, trunk, and branch. Then only maintenance work and laying by of stored food goes on, and even this stops when the leaves begin to turn color.

In the quiescent state known as rest or dormancy the aboveground parts of hardy plants, whether trees, shrubs or herbs, are highly resistant to cold. Dormancy sets in during the summer, gradually intensifies to a peak in early winter, and then subsides, so that by late winter it is no longer internal forces but rather environmental cold that holds the dormant buds in check. Early in the "rest period" almost any kind of physiological jolt will rouse the buds and set off a new flush of growth. Treatment with heat or cold or any number of chemicals, sometimes even heavy watering or a generous dose of fertilizer about the roots will serve. By Christmastime, however, rest is ordinarily deep, and very few kinds of buds can be forced into growth by bringing them into a warm room. Then as winter goes on, the different kinds of plants one by one become ready to start growing again whenever conditions around them are suitable.

To break its "rest" so that it can resume active growth, each kind of hardy plant has its own particular "chilling requirement." Both the intensity and the duration of cold are important, although contrary to popular belief, the most effective temperatures are not necessarily below freezing. The low forties will do in most cases; and even a mild winter on Cape Cod, for instance, is cold enough to meet the needs of the most demanding plants.

One might think that plants native to the coldest regions would require the greatest cold to break their dormancy. This, however, is not the case. It is true that some southerners such as almonds and pecans need little exposure to cold; but some plants from the most rigorous winter climates also need only moderate chilling. Keeping dormant during the depth of a cold winter is no problem; the cold external environment will effectively restrain the activities of any eager little wakening buds. It is the long, mild autumns of the middle regions and the occasional warm periods in midwinter that necessitate some sort of brake on the expanding of winter buds.

Summer or winter, in its primeval state the northern forest was a place of great, dark trees overhead and damp, decaying humus underfoot. The last tract of virgin forest to survive in Connecticut, in the town of Colebrook in Litchfield County, was a beautiful stand of hemlock and northern hardwoods. Before lumbering was begun there in 1912, the beeches and hemlocks that accounted for over half of the individual trees were anything up to 350 years old, some a hundred feet tall with trunks a yard through and rising a clear fifty feet from the ground to the lowest branches. Among these grew many sugar maples and yellow birches, with scattered trees of half-a-dozen other kinds such as oaks and ashes. Young seedlings and saplings were of the same kinds as mature trees, showing that this was a climax forest that would perpetuate itself here indefinitely.

Among the hemlocks and hardwoods of New England's primeval

forest white pines stood out conspicuously because of their dark color and especially because of their great size. Where the general height of the forest might be eighty to a hundred feet, the pines, singly or in groups, towered more than a hundred and fifty feet tall. In some places there were large stands of almost pure pine.

The best of the great pines were always the first trees to be cut for lumber. By the time the virgin pines were seriously depleted, pure stands of a new though inferior crop were growing up on large tracts of abandoned farmland. When the old-field pines had been cut and another generation of woodland began to develop in their wake, New Englanders were in for a surprise. Common as white pine has always been throughout the region, it does not ordinarily replace itself when an old stand is once removed. Even in a wholly natural wild situation, if the soil is passably fertile the young trees that start in the underbrush of an old pinewood are not more pines, but hemlocks and hardwoods. Only where poor, sandy soil stunts the growth of the more rampant hardwoods can young pines establish themselves and keep the upper hand.

In view of all this, the role of white pine in the native forest was for many years an ecological puzzle. Clearly it is not just crowding for living space that is involved. Pine seedlings can grow in pasture sod that is much more densely populated than a forest floor. It has taken long and careful study to learn that the specific critical factor is light. Young white pines will not tolerate any serious degree of shading, and can grow vigorously only in the open. In that case one may wonder why they were always present in the primeval forest.

The answer is that a large virgin forest is by no means a wholly unchanging world. Openings are constantly being made in the overhead canopy, large ones by fire or severe windstorm, smaller ones by smaller blowdowns or by the death of single trees from blight or lightning. A large stand of virgin pine probably marks the occurrence long ago of a big storm or a fire severe enough to make a complete kill and let in full daylight over a wide area. In such stands the trees are commonly all the same age, indicating that they started during a single season when conditions were just right for large numbers of seedlings to become established in a place free from overhead shade. A smaller opening would make space for fewer seedlings, and of these only a handful or even a lone one might survive to maturity. Isolated pines scattered within a small area may be of widely differing ages, showing that they all started at different times, each as the result of a small, isolated incident in the forest. It seems, then, that though white pine is a permanent part of the north woods at large, in any given spot it has always been a transient.

Beneath the tallest trees of the northern forest grows a characteristic lesser vegetation. Among the smaller trees and shrubs are striped and mountain maples, mountain ash, witch-hazel, and several of the horde of species of viburnum. Small herbaceous plants growing close to the soil are largely different from those of the oak forest. Where there are not too many evergreens and the forest floor is relatively bright and warm in early spring, there are patches of clintonia, woodsorrel, starflower, and goldthread. There may be carpets of bunchberry, like miniature dogwood, white with blossoms in May and June, red with clusters of berries in late summer. On peaty tussocks or rotting stumps one may find the fragrant little pink twinflower, *Linnaea borealis*, that grows in north woods around the globe. Ladyslippers and other orchids can be found, although nowhere in great abundance. Where many evergreens cast a year-round shade on the forest floor, flowering plants are scarcer. Instead there is an abundance of ferns and mosses of many kinds. With good fortune one may come upon tawny beechdrops or the deadwhite Indian pipe, that live not by their own efforts but by drawing on the roots and fallen leaves of large trees.

Beneath the virgin forest the soil, although well-drained, was covered with a moist, springy cushion of humus as much as a foot deep. Lying on this were ridges and mounds made by windthrown or otherwise fallen trunks in all stages of decay. In the primeval state the soil, the logs and stumps, even the trunks of standing trees were plastered over with a thick, moist blanket of mosses and liverworts. Not even the best of our second-growth forests have attained to such a condition. The better soils have only a thin carpet of leafmold, and only cool ravines and the margins of woodland brooks are damp enough to allow the growth of a mossy blanket.

The soil in the forest is in a constant state of flux. Wherever a large tree blows down, its dislodged roots tear a hole in the earth, exposing a little patch of mineral soil that can serve as a seedbed for infant pines and others that prefer such a place for germination. As a tree goes over, part of the mass of soil and stones that has become entangled with its roots is picked up and deposited on top of the humusy forest floor next to the hollow where the roots gave way. The formation of windthrow mounds and their adjacent hollows in this manner brings about a slow but continual mixing of the forest soil, burying masses of humus and depositing layers of subsoil on the surface. This in time produces a highly fertile, mellow soil that furthers the growth of the forest trees. The process still goes on, and in present day forests the paired mounds and hollows can be found in all stages from freshly thrown trees to scarcely perceptible bulges on the surface of the soil.

In the north woods as in the oak forest, vegetation is closely interrelated with the underlying soil. Good soil supports broad-leaved species, and these in turn improve the soil they grow on. Under a pure stand of evergreens, on the other hand, the needles, like oak leaves, accumulate in a thick layer that decays extremely slowly and remains sharply distinct from the mineral soil below.

For several inches just beneath the humus mat the soil may be a startling ashy white. In this layer constant washing by trickling rain water dissolves away a large fraction of the mineral salts and carries them, along with any dark humus particles, deeper into the soil. There is very little replacement from the slowly decaying surface litter; so the upper mineral soil becomes a barren, whitish sand, much impoverished and highly acid. The ashen appearance of this layer is responsible for the name "podzol," given to this kind of soil by Russian scientists, who were the first to study soils in an intensive and systematic way and whose names for the different soil types have become standard technical terms.

Real podzol soils are found only in the far north and high on mountainsides, mostly in the spruce-fir country. A true podzol is good for very little except growing coniferous trees. South of this and at lower elevations lies a large zone where the same processes operate to a less extreme degree, giving rise to what are called "podzolic" soils. With good management these can be greatly improved and even made productive for agriculture, if other local conditions warrant the effort.

Within the northwoods area of New England, more or less correlated with soil as well as with climate, one can recognize two subdivisions according to whether white pine or spruce is more abundant. Here again the distinction is perhaps easiest to see in old pastures. From central Maine northward into Canada, eastward along the coast and westward on the higher lands, the common pasture invader is not white pine, but spruce, sometimes accompanied by balsam fir. Balsam behaves much like spruce, but is shorter lived and less tolerant of shade. Where the soil is better and exposure less severe, spruce is mixed with the same northern hardwoods that grow with white pine, typically sugar maple, beech, and birch.

Spruce in its own region seems to have a status like that of white pine farther south. Both come in as pioneer trees after the forest has been disturbed and persist against hardwoods only under adverse circumstances. Where perpetual winds blow in from the ocean, or high in the mountains, spruce ventures farther out than any other trees. Where the soil is shallow and stony, sometimes where there is no real soil at all but only a layer of organic matter lying directly on bedrock, spruce may be the only tree that can grow.

However lonely and wild the north woods may look, practically none of it in New England is virgin forest. There are some "old growth" tracts, where large trees make an impressive forest. But the size of trees may be deceiving. In parts of Maine where growing conditions are good, white pine stumps thirty-two inches across have been found to average only seventy years old. Even in the far North and in mountain fastnesses, practically all the forest is growing on land that has at some time in its history been clear-cut for lumber, or even farmed.

Lumbering has gone on without interruption from the days of the Pilgrim Fathers. The first sawmill in the country was sent over from England and set up near South Berwick, Maine, in 1634. Logs were probably cut on Mount Desert Island long before the first permanent settlement there in 1762, and large numbers of them were rafted down the Connecticut River from New Hampshire in the 1730's.

In the early days the forest seemed endless and inexhaustible. Plenty of timber was available near the settlements, and trees were cut only for local consumption. As the country built up and loggers began to branch out, they at first took only the best pine. For a long time cutting was limited to places that were accessible by water, where lumber could be transported by ship, or where at least the largest logs could be floated down river to the sawmills. For years during the period of agricultural settlement transportation was so difficult that only the choicest timber had any commercial value at all. The rest was simply gotten out of the way by girdling and burning.

The advent of railroads beginning in the 1840's brought a great change to the north woods. A vast new territory was opened for lumbering to meet the clamorous demands of a rapidly expanding young nation. There followed a period of reckless cutting of forests, the best trees first, but eventually virtually everything that could be sawed into boards, and never mind what happened to anything that couldn't. Between cutting for lumber and clearing for farms, about the only virgin forest left by the turn of the century was in remote regions such as the Rangeley Lakes in Maine; and in the 1890's the railroad penetrated even there.

In the past half century, methods of exploiting the forest have changed once more. When the good virgin timber was gone, the great lumber camps moved on to the west; but by then a new generation of young trees was beginning to appear on cutover land and abandoned farms in the east. In central New England this grew up to be old-field pine, which was harvested over a period of roughly fifty years from about 1880.

In the north, the pulpwood industry got under way at about the same time, making use of leftovers and second growth wood. At first the pulp mills could use only spruce; but paper chemists were hard at work and soon devised methods for using balsam fir. By 1912 even hardwoods could be used in the pulping process. Around 1930 motor trucks invaded the woods, and now it is possible to both harvest and use all kinds of trees from all but the most outrageously inaccessible places.

Driving through the wooded north country today one sees evidence of widespread cutting of pulpwood. There are piles of small logs by the roadside, and truckloads of them in transit. Logs are still driven down the rivers to a certain extent, or more often towed in great rafts held together with chains. Larger rivers are used for storage as well as transport. For miles above a paper mill the water may be covered with a solid pavement of floating logs, kept in place by long boom chains attached to rock-filled cribs in the river. At the mill the logs are taken from the water, carried up a sort of escalator conveyor, and thrown like matchsticks onto the top of a mountainous stockpile.

Logs large enough to saw into boards and timbers are seen far less frequently, although a certain amount of lumber is still produced in New England and used locally, much of it for making furniture. Ways are being sought to co-ordinate the activities of furniture manufacturers with those of paper makers so that logs can be sorted at a common mill and each one used for whichever purpose it suits better; but this is at present a more hopeful than practical proposition.

Apart from its long history of extensive cutting, the forest shows the marks of repeated burning. In the early days of clear-cut logging, tops and trimmings—"slash," in loggers' parlance—were simply left where they fell. Sometimes the slash rotted away eventually and nothing dreadful came of it. But if fire once started in dry twigs and needles, the roaring heat was enough to burn not only such plants as remained, but also the entire accumulated mat of organic matter. With the humus gone, revegetation had to start on impoverished soil or often on bare rock. Heavy rains in the interim would make serious erosional inroads into what soil remained.

Fire has been brought under much better control in recent years, but a severe drought made possible the holocaust of 1947 in Maine. Ten years later the country around Bar Harbor still shows clearly the raw, unhealed scars of that catastrophe.

Even far less severe fires may kill susceptible species and injure parts of the more resistent plants. When the above-ground part of a tree is killed, the consequences depend on the regenerative powers of its root and stump. None of the northeastern evergreens except pitch pine are capable of sprouting, though oaks and maples can, and birch and aspen sprout profusely. What happens after a fire, then, will depend partly on the nature of the pre-fire forest.

Predominantly hardwood stands may replace themselves with relatively little change except for an increase in the proportion of the more rapid growers. In a mixed forest, fire will eliminate the conifers, leaving only hardwoods. An evergreen forest will be replaced by sprouts from any scattered hardwoods that survive and more especially by light-seeded species such as birch and aspen.

The birches that are so attractive in the north country often grow in almost pure, even-aged stands. They may occasionally mark the site of abandoned farmland, but much more commonly they come in after fire. The species is a clue to the nature of the soil. Gray birch is a sign of dry, poor, or badly abused land; white birch indicates better soil conditions where the original forest was predominantly hardwood and, given time, will be so again. Though one may admire an open, airy birch grove extravagantly, it is nevertheless a sign of some past forest calamity.

In the spruce country, fire may be followed by a generation of birch, which may have to grow to considerable size before enough humus accumulates to make a good seedbed where more spruce can get started. In an area where no trees at all will grow except spruce, it may be years before the land will support much more than fireweed and bracken.

Truly virgin forest is hard to find nowadays, but there are relatively undisturbed places where one can at least see what it consisted of and taste its atmosphere. In western Connecticut near the village of Cornwall there is a grove of huge trees known as the Cathedral Pines. It is not certain whether these are virgin timber or a very ancient old-field stand. The Mianus River Gorge near Greenwich harbors a virtually primeval stand of beech-maple-hemlock forest. In western Massachusetts the lower parts of the Mount Graylock Reservation and Savoy Mountain State Forest are covered with hemlock and northern hardwoods, and their peaks reach up into the spruce zone. Smugglers' Notch near Stowe, Vermont, and the steeper notches of the much-disturbed White Mountains still have fair samples of northern hardwoods. The Jefferson Notch road goes through spruce type forest. Farther afield there is Baxter State Park in north central Maine, where the vegetation ranges from hemlock and hardwoods on the lower levels to treeless alpine tundra on the top of Mount Katahdin. Everyone will have his own favorite; but for a bit of country that is at once easy of access and full of sweet, northern wildness, I will go every time along the road that follows up the Connecticut River to its source near the Canadian border.

10

The Mountains

NEW ENGLAND's hills and mountains are of very ancient lineage. The geological convulsions that gave rise to their original ancestors took place a long time ago even in terms of geological ages. What we see today are merely the exposed roots of mountains whose upper bulk has long since been stripped away by erosion through an almost inconceivably long reach of time. The shifting and heaving of the earth that went into the making of the ancestral mountains and the great pressures that acted on what then were basement layers have folded and broken all but the youngest rocks almost beyond deciphering. The general course of events is not too hard to understand, but when it comes to ancient details, a geologist may prefer to talk of other things, and he sends his students to the newer Rockies, where the workings of the earth seem by contrast clear and simple.

Whatever the one-time alpine scenery of New England, a large part of its surface was subsequently eroded almost to flatness. There have been minor ups and downs of the earth's crust meanwhile, but no real mountain making has gone on for several aeons. The present roughness of the country has been wrought not by upheavals of the earth, but rather by the less dramatic forces of erosion acting on a relatively stable land.

The uplands of New England fall rather clearly into two categories according to their erosional history. The higher peaks that qualify as mountains in this part of the world are remnants of the primordial alps that were never quite smoothed away by moving water or ice. Some of them survive because they are made of especially resistant rocks. Others lie on inland watersheds that the headwater streams have not yet had time to level off. The lower hills that form a large expanse of rolling upland are the knobs and ridges left between the many valleys that have been carved into a once flat land surface. The old, smooth plain can still be seen with only a small effort of imagination wherever there is a wide view across the low hills. Then it is apparent that all the hilltops for long distances around rise to about the same height; and a line connecting their crests makes an even skyline that represents the one-time level land surface.

The higher mountains of New England fall into several quite different groups. Whether or not one extends the term "mountain" to include the Berkshires is a matter of definition. Their official name is the Berkshire Hills. This little rough plateau land is a clearly bounded entity, its eastern side rising abruptly from the Connecticut Valley, its western dropping sharply into the Hoosic and Housatonic Valleys. The northern part of the Berkshires is really the trailing end of Vermont's Green Mountains; their southern end tapers off into the hills of northwestern Connecticut.

The two thousand foot elevation of the Berkshires is not an impressive altitude in an automobile in summertime. But a traveler crossing over it during the changing seasons of spring and fall finds the weather and the state of the vegetation quite different on the hills of West Cummington from what he left half an hour ago in the valley at Northampton. In April, for example, when crocuses are out and tree buds bursting in the valley, the hills are still locked in wintry snow and ice. Not only is the year-round climate colder on the upland, but the growing season—the time between spring and autumn frosts—is substantially shorter. Botanists figure that for plants, a thousand feet up is the equivalent of six hundred miles or

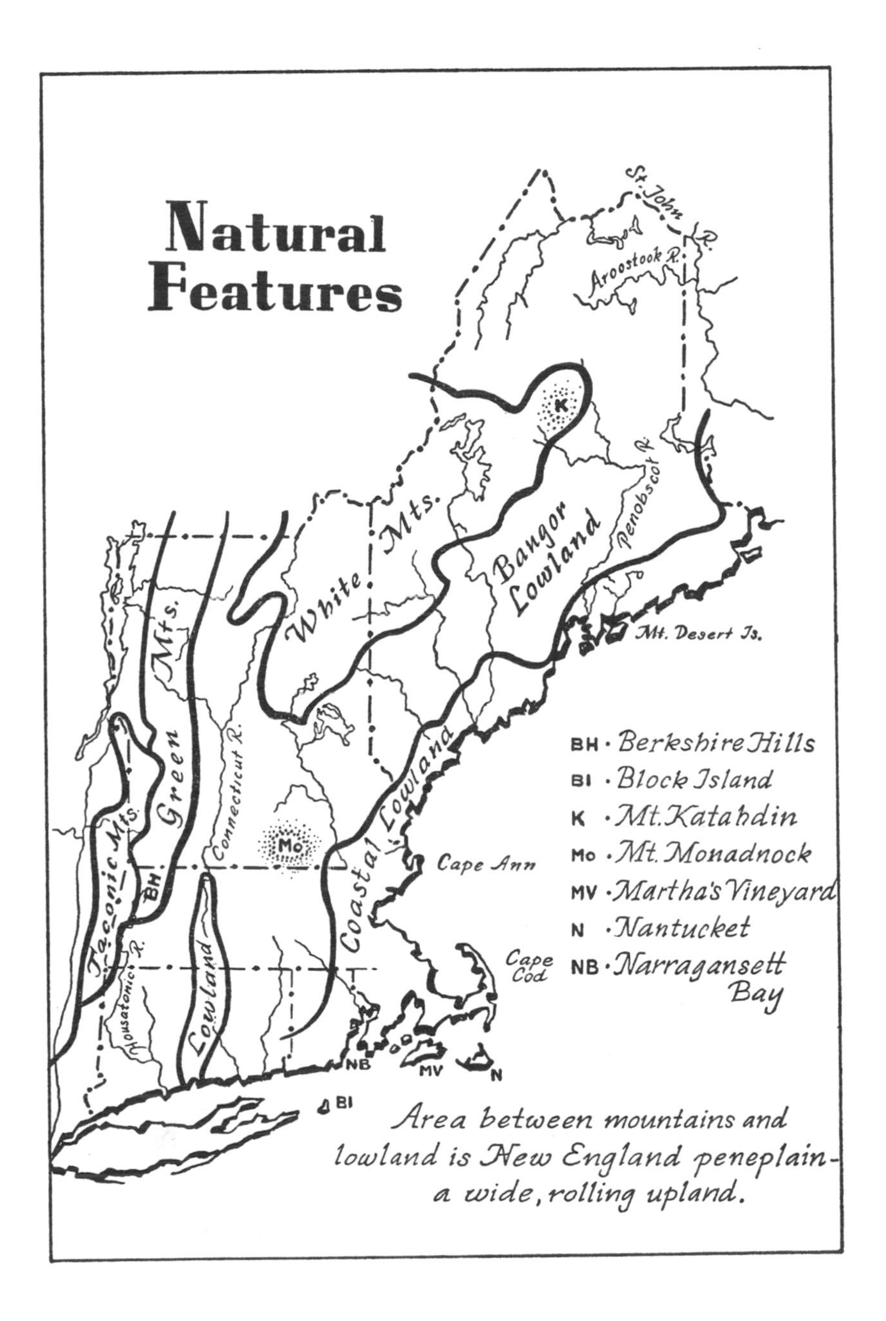
Natural Features
St. John R.
Aroostook R.
Penobscot R.
K
White Mts.
Bangor Lowland
Mt. Desert Is.
Green Mts.
Connecticut R.
Taconic Mts.
BH
Mo
Coastal Lowland
Cape Ann
Cape Cod
Housatonic R.
Lowland
NB
MV
N
BI
BH · Berkshire Hills
BI · Block Island
K · Mt. Katahdin
Mo · Mt. Monadnock
MV · Martha's Vineyard
N · Nantucket
NB · Narragansett Bay
Area between mountains and lowland is New England peneplain - a wide, rolling upland.

more north; indeed, fields and forests high in the Berkshires have a quite northern aspect compared to that of the nearby valleys.

In colonial times these modest hills that we find so charming today were a serious hindrance to pioneer settlement, so much so that they were long called the "Berkshire Barrier." The first white men who crossed them, traveling on foot and on horseback from Westfield in 1694, reported that "the greatest part of our road this day was a hideous howling wilderness" beset with "hideous high mountains." This was clearly a sentiment of the days before the romantic age revolutionized our attitudes toward scenery.

Pioneer settlers finally pushed into the Berkshires in the 1720's along the old trail followed by the present road from Westfield to Great Barrington. Once settlement began, the hill country developed rapidly. It reached its vigorous heyday about the time of the Revolution and continued for several decades in this happy state. During that time the entire primeval forest was either removed or at least drastically altered by clearing for farms, logging for sawmills and tanneries, and making charcoal to feed the local iron foundries. Then around the time of the Civil War the great rural decline that affected so much of New England's hill country struck hard at the Berkshires; and here, too, the forest has since come back.

The flavor of the changing times can still be sensed in contrasts of the present day. To see it, visit first the boyhood home of William Cullen Bryant. He was born at Cummington in 1794, the son of a frontier doctor. The house has been preserved, a substantial dwelling that still looks out over open, rolling fields. Though the countryside is quieter now, one can easily in imagination picture it checkered with many small, flourishing farms, and reconstruct the village as a busy little place with a surprisingly large number of small industries, from harness makers to textile mills and spool factories.

Then go to Buckland and search out Mary Lyon's birthplace. When she set out on the path that finally led her to the founding of Mount Holyoke Female Seminary in 1837, she came down from a simple but comfortable farm home in a well populated rural district. There were many neighboring houses in sight here and there across the open fields. Today you follow a back road up into the lonely, wooded hills, and find at journey's end only a cellar hole in a small clearing, the spot marked by a bronze plaque on one of the ever-present boulders.

Only those versed in geography or geology remember that the Taconics are a separate range of mountains. They stretch in a long, narrow belt along the line where New York State adjoins Vermont and Massachusetts, sweeping sharply up on the west side of the great limestone valleys of the Housatonic and the Hoosic and the marble vales of the Batten Kill and Otter Creek in Vermont. The range is studded with monadnocks, highest among them Graylock, Equinox, and Dorset, all rising to about 3,500 feet. This is a modest altitude, but it has its full dramatic value in contrast with the broad, deep valleys from which the mountains rise. Mount Graylock is an imposing mass when seen from the six hundred foot valley floor only a few miles away.

The Green Mountains virtually *are* Vermont. These ridges with their parallel valleys that run the length of the state are the last remnant, the very core, of an ancient up-fold of the earth's crust. In the southern part of this venerable wrinkle, thousands of feet of overlying rock have been stripped away, and the ridge of the mountain crest exposes pre-Cambrian rock. This is the oldest kind on the face of the earth, so old that no surely recognizable fossil of even the lowliest of once-living creatures has been found in it.

Much of the rock that forms the Green Mountains consists of gneiss, a hard and resistant type, rather like granite, that has been sheared and compressed in such a way that it has taken on a banded structure. The bands were once straight and level, and their present contorted pattern is evidence of the intense bending and folding the rocks have undergone in their long history. In the Green Mountains folding has been so strong that in some places, especially toward the north, the folds of rock have broken and toppled over on their sides. Since then the upper layers have been pushed westward for as much as fifty miles, overriding the younger layers of the former surface like a giant sledge. Where banded or layered rocks are exposed in a cross-sectional view, one can see for himself their folded and crushed nature and imagine the intricate processes that have gone on through the long mists of geological time.

Across the river in New Hampshire lie the White Mountains. Their tallest peaks form the Presidential Range, where Mount Washington rises 6,290 feet above sea level, the highest point east of the Rockies and north of the Smokies. The White Mountains lie in clumps and clusters that lack the clearly elongated ridge and valley pattern so common in eastern America. In the broader sense they include several groups that straggle eastward for a hundred and fifty miles to Mount Katahdin, a 5,200-foot monadnock outlier in northern Maine. All of these mountains are geologically related, although several groups among them have their own names, like the Blue Mountains and Boundary Mountains.

West of Katahdin and north of Moosehead Lake lies a large expanse of rolling plateau land that stands some 1,100 to 1,250 feet high, covered with a forest of spruce and hardwoods and strewn with lakes and bogs. So few people travel this remote wilderness that its great extent is not commonly realized. Then in extreme eastern Maine the plateau falls away to a lowland formed by the upper valleys of the Penobscot, the Aroostook, and the Saint John Rivers. Except for the potato region, this is still a forest country, and in spite of more than a century's lumbering, the whole area still answers quite well to Thoreau's description of it in "The Maine Woods."

The White Mountains proper, in New Hampshire, are New England's most mountainous mountains. They were born long ago when a tremendous mass of molten rock welled up from deep within the earth and eventually solidified into a dome of tough, resistant granite. The thousands of feet of softer rock that originally overlay the granite have long since been worn away. The peaks we know were sculptured from the remaining hard core by streams whose drainage pattern still radiates from the center of the mass. Rock structure here has little influence on topography, except in a few places such as the Flume Gorge, where a mountain stream has cut deep down along a sheet of softer basalt that strikes through the general granite.

When the climatic changes began that in time produced the continental ice sheet, small glaciers formed in the highest mountain valleys, especially on the east side of the Presidential Range. These became large enough and lasted long enough to gouge their valley heads into steep, amphitheaterlike glacial cirques. Perhaps the most familiar one is the head of Tuckerman's Ravine, where in the chilly shadows snow lies late in spring, to the joy of ardent skiers. Other ravines and "gulfs" originated in the same way. Lower down, the rivers of ice scoured their sharp, narrow valleys into rounded troughs with broad, smooth bottoms like that of Crawford Notch. All these features are rudimentary but classic examples of true alpine geology.

The separate valley glaciers had no more than begun their work and had scarcely nibbled into the mountain mass when they were swallowed up in the advancing continental ice sheet. Eventually the ice flowed over even the highest peaks, scraping and smoothing the contours of the solid granite. In spite of the extreme weathering that takes place on the heights, stray erratic boulders dropped by the glacier remain almost on the very crests as tokens of the passing ice.

The glacier, when it finally melted, left behind its usual mantle of bouldery till. This still remains on the lower slopes, but higher up little but the larger boulders has withstood the wash of running water, and soil becomes increasingly scarce. Farther down, the broader valleys among the mountains are spread with a layer of glacial till that makes a soil more or less adequate for farming, especially in the spots that once were occupied by temporary post-glacial lakes. Such relatively level areas among the mountains are locally known as "intervales." No very serious farming is attempted in the higher valleys, but some of the intervale meadows still produce a passable crop of hay.

The vegetation that covers the mountains is neatly zoned according to the altitude and exposure where it grows. A journey up Mount Washington takes one through several different kinds of forest zone and finally, at the top, out into the open above timberline.

The feet and lower slopes of the mountain are clothed with a forest of hardwoods—beech, maple, birch, and others—mixed with red spruce and balsam. Farther up, some of the hardwood species begin to drop out, leaving white birch and mountain ash relatively more important. As the cold and the wind increase, black spruce appears, first as vigorous, symmetrical trees, then more and more showing the effects of the struggle with the elements. At first the trees are merely stunted because of their extremely slow growth in the cool air and stone-bound mountain soil. Then as the force of the wind becomes increasingly a factor to be reckoned with, the trees become increasingly deformed, although in sheltered hollows or in the lee of rocks or ridges, needles and twigs are normally formed and thrifty in their growth.

The shearing effect of the wind can be seen everywhere. A few upright trees stream away from the prevailing wind in a flag form. Other venturesome stalks rising above the general surface may survive for a few years; but eventually they succumb to the fury of a severe winter, leaving their bleached skeletons as monuments to foolhardiness. Plants that cannot grow upward will then grow horizontally. The windshorn spruce sends out a multitude of low, ground-hugging branches that develop into a tough, impenetrable tangle firm enough to support a man walking over the top. The smoothly pruned contours of the matted thickets show just where the edge of the wind rushes past, and where the winter snow collects in sheltered and sheltering drifts or is swept clean away by the scouring blast.

Still higher, where not even deformed and stunted spruce can survive, dwarf shrubby species of birch and alpine willows appear. These also grow in a dense mat, never more than knee-high, and commonly reaching scarcely to the shoetops.

At last you come out above even the dwarf trees onto true alpine tundra at the top of the world. On a clear day there is nothing to break the view for a thousand miles or more except the earth's own curvature. To stand as high as the peak of Mount Washington you would have to go to Mount Mitchell in North Carolina or the Black Hills of South Dakota. The wind blows in a steady torrent, and much of the time it carries wisps and drifts of cloud through your streaming hair. When you take shelter from the wind's whistle, you can hear the songs of distant hermit thrushes and white-throated sparrows floating up from the forest far below.

All around the earth falls away in a rolling sea of gray rock slabs and boulders. Loose soil has long ago been washed or blown from the exposed places and filtered down into the crevices, leaving the mountain top covered with a rough, cobbled pavement of giant stones.

Here on the tundra in the nooks where anything at all can grow, the most abundant plants are grassy-looking little sedges. More picturesque are the evergreen mats or cushions of species that grow in the far North, as their scientific descriptive names of *lapponicum* or *groenlandicum* reveal. So far as plants are concerned, a trip to the top of Mount Washington is the equivalent of a journey of hundreds of miles, as far north as Labrador or the Arctic Barrens.

Alpine flowers, the rarer the better, have always held a fascination for lowland dwellers. Specimens of all of these have been collected, identified, and cherished by countless amateur botanists through the years. They are indeed charmers, neat little mounds or doilies of small, dark leaves, usually evergreen, thickly studded in June and July with flowers of the most fetching forms and colors. There is the Lapland rosebay, *Rhododendron lapponicum*, like small red-purple azaleas. There are *Diapensia lapponicum* and the smaller mountain sandwort, *Arenaria groenlandica*, each one a cushion of dainty white flowers. Tiniest of all is the minute alpine azalea, *Loiseluria procumbens*, with bright pink blossoms scarcely a quarter of an inch across. Less prettily colorful but equally exotic as northern tundra plants are crowberries, mountain cranberries, and prostrate dwarf willows. There are many others. Some of them grow on other high peaks in New England and the Adirondacks, but to see them all in their northern lowland homes one would have to cross hundreds of miles of intervening forest where not a one of them is to be found.

People familiar with our western mountains are not likely to be impressed with the mountainous character of five or six thousand foot peaks. Even a novice, provided he is strong enough in lung and limb, can climb any of the trails on Mount Washington. Then, too, the whole region has been swarming with tourists almost since it was first settled in the 1770's. The most sedentary of travelers can reach the top of the highest peak, taking his choice of auto road or cog railway. Yet Mount Washington has taken far more lives than any other mountain in America, and those who have worked both here and in the Antarctic say that the top of this little mountain has the worst weather in the world.

Consider for a moment what the place is like. Winds of hurricane force occur here in every month of the year, in winter on an average of four days a week. Two-hundred-mile winds blow occasionally, and the highest wind velocity ever measured, 231 miles an hour, was recorded here. In January the wind roars past at an average rate of fifty-five miles an hour. In July, even counting in the occasional still days, the wind speed averages a smart twenty-three miles an hour.

Temperatures are also severe and show strongly the effect of altitude. Not far away in Boston, winters vary around 30° F. On Mount Washington the figure is 7.1°. Whereas summer averages 72° on the coast, it is 48° on the mountain. Once the mercury fell to 49° below zero. And it usually snows sometime during every month of the year.

These are the conditions that mountaintop plants have to contend with. A man on foot must take into account not only wind and cold, but also blinding rain and mist. Here are a few more figures. It rains or snows on 57 percent of the days in an average summer, and half the mornings in July are foggy until well past sunup. The close spacing of stone marker cairns along the rocky foot trails near the summit should remind the fairweather hiker of the very real hazards of this extremely exposed place. If this is not graphic enough, there is a hair-raising little book called "Three Days on the White Mountains, being the Perilous Adventure of Dr. B. L. Ball on Mount Washington during October 25, 26 and 27, 1855."

Scientists have wondered why timberline on New England mountains should lie at about half the elevation that it does at the same latitude in the west. Probably it is the completely exposed position of the eastern peaks that accounts for the difference. There is nothing to break the force of the weather, nothing within a thousand feet of the height of Mount Washington and very little approaching that between it and the far side of the Great Plains.

Plants exposed to the eternal wind that blows here must survive not only its mechanical force and its cold and drying effects, but also the onslaught of wind-driven snow and sleet. Typical alpine plants hug the ground or lie low in pockets and crevices between the rocks. In this way they really avoid rather than endure the effect of the wind. In larger sheltered spots and hollows, especially where protecting snowdrifts lie all winter, trees can grow right up to the ceiling that the wind makes. In 1940 a visiting botanist was able to find small, isolated trees growing right to the mountain top, the highest one a fir standing not more than seventy feet from the summit. All this strongly suggests that it is not so much extreme cold as wind that determines the location of timberline on our eastern mountains.

Alpine tundra above timberline closely resembles that of the Arctic. On Mount Washington alpine plants are especially abundant on the relatively level, almost lawnlike expanses long known as the Alpine Gardens. Like the arctic tundra, these are cold, wet, windy places where the soil is stony and sparse. Here the earth is well drained because of the topography, but always moist from the frequent rain and mist. Much of the area is a jumble of rock slabs, with soil enough for rooted plants only in small pockets in between. Those who have tried digging report that there is more good soil down under than is commonly believed; but overlying rocks conceal most of it and make it inaccessible for small plants.

The surface soil where small plants are rooted is in a constant state of flux. For a large part of the year it alternately freezes and thaws. This has a churning effect and makes the soil do a sort of creep movement down hill. Even level spots where soil is a little more abundant are subject to constant stirring. In addition to this kind of disturbance, high winds from time to time dislodge and blow away tufts of vegetation along with the soil they are growing in. People familiar with the high country report instances of coming back repeatedly to a favorite clump of some little flower until one year it just isn't there any more.

In spite of years of botanizing by many people on the higher mountains, little is known for certain about the relationships of the plants that grow there to their environment and to each other. Wind and cold and dampness are all obviously important, as well as the instability of the soil. But which of these factors is critical? Is it the same one for all plants? Is it merely chance that brings the various alpine species into now one and now a quite different combination? For these and many more questions there are more or less informed guesses but no well-founded answers, and some hardy mountain lover with a bent for experiment and a knowledge of plants and of measuring instruments has a wide open field for research.

The treeless area on Mount Washington is some eight miles long by two or three miles wide. There are similar but smaller places on no less than twenty-five of New Hampshire's peaks, as well as on Camel's Hump and Mount Mansfield in Vermont and Katahdin in Maine. Most of these can be reached only on foot. Happily for those with willing spirits but weak or rebellious flesh, it is possible to journey to the tundra of Washington or Mansfield in an hour's ride in the ease and comfort of one's own automobile.

II

The Connecticut Valley

CLOSE TO THE Canadian border in the northernmost end of New Hampshire lies a little brown pond perhaps six feet deep and two and a half acres in extent. This is the Fourth Connecticut Lake, and the trickle leading out of it is the infant Connecticut River. It has not very far to go to reach Third Lake, and there the traveler by road can make its acquaintance.

Coming up from the New Hampshire town of Lancaster, U.S. Highway 3 follows along the young Connecticut toward Canada. Just before reaching the border it skirts the east side of Third Lake, almost touching the shore in places. Here a few steps from the road bring one out on a narrow sandy beach at the edge of a true north woods lake. The water is clear and cold, as much as a hundred feet deep in one spot, and fringed with a dark forest of spruce and fir. In the background here and there rise the blue tips of distant mountains. On summer weekends there will be a few fishermen for company, but at other times one may have the place quite to himself.

The road leading south, downstream, winds off into rolling forest-land that has been little disturbed except for periodic lumbering, and of course and alas! an occasional fire. The first fifteen miles of the journey toward First Lake is a succession of north woods delights. There are dark forest and dappled glades, broken here and there by

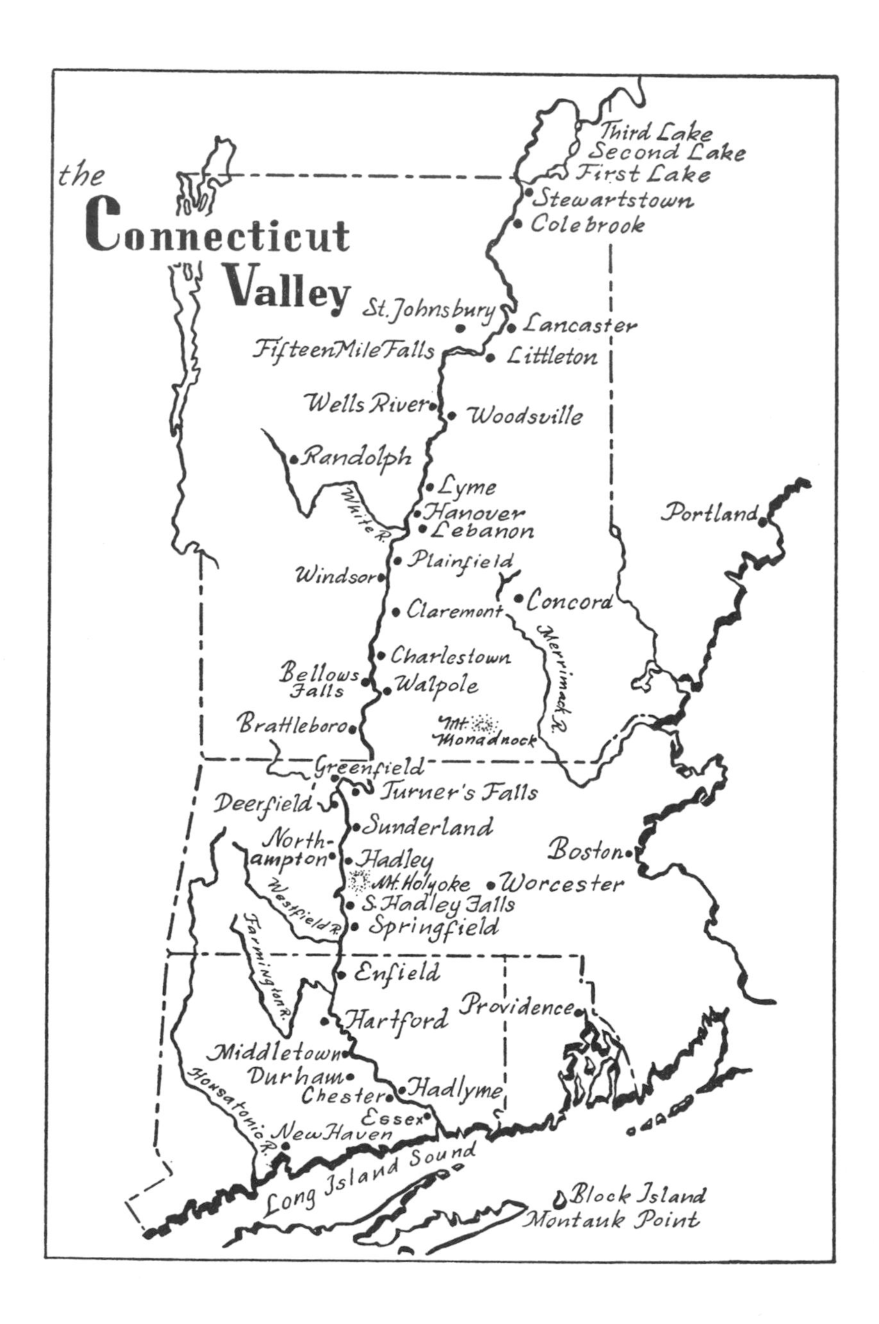

the
Connecticut
Valley
Third Lake
Second Lake
First Lake
Stewartstown
Colebrook
St. Johnsbury
Lancaster
Fifteen Mile Falls
Littleton
Wells River
Woodsville
Randolph
White R.
Lyme
Hanover
Lebanon
Portland
Plainfield
Windsor
Concord
Claremont
Merrimack R.
Charlestown
Bellows Falls
Walpole
Mt. Monadnock
Brattleboro
Greenfield
Turner's Falls
Deerfield
Sunderland
North-ampton
Hadley
Boston
Mt. Holyoke
Worcester
Westfield R.
S. Hadley Falls
Springfield
Farmington R.
Enfield
Hartford
Providence
Middletown
Durham
Chester
Hadlyme
Housatonic R.
Essex
New Haven
Long Island Sound
Block Island
Montauk Point

sunny meadows and swamps. Strawberries grow in the grass at the roadside. Twinflowers trail over crumbling stumps in the shady edge of the woods. There are dark spruce groves where little can grow in the thick mat of fallen needles. And now and again one catches a glimpse of the young river, almost hidden in an alder thicket or widening out to sun itself broadly in a shallow, golden-brown riffle.

From First Lake on down, the influence of man's activities becomes increasingly strong, beginning with sporting camps, open, sunny hay meadows, and a dam at the lake's outlet for the benefit of logging and power requirements; but in all the valley's length only a few spots are marred by industrial ugliness, and everywhere the river lays its beauty on the land.

The valley where the river flows has existed in one form or another through the entire long course of New England's geological history. The story begins far back in Cambrian time, five hundred million years ago, with a long chain of low, mountainous islands rising from the sea to form the ancestors of our Berkshires and Green Mountains. Later, but still a long time ago, land farther toward the east rose to form another range of ancestral mountains. And along the trough between, the primordial valley came into existence.

Since then the land has been uplifted and depressed several times and compressed into titanic folds. Always the buckling and cracking of the rocks has followed a north-south axis, so that the entire region has a strong lengthwise grain. Sometimes this has been apparent in the trend of mountain ranges. Sometimes it has been largely concealed when the mountains were worn away and the land eroded almost to flatness. Even then the grain was there, exerting its influence on the courses of rivers flowing over it.

There was a time, while the face of New England was at its smoothest and flattest, when the entire region became tilted upward on the west, so that the main rivers of the day flowed southeastward to the sea. As the streams cut down into the land, the underlying north-south grain of the rocks began to make its presence felt.

Wherever a small tributary river chanced to unearth a zone of less resistant rock, it rapidly outgrew its main stream. To the present day our largest rivers flow mainly southward along deep soft-rock valleys.

In a sense the direct ancestor of the Connecticut River that we know today is the little Farmington River. It was a small branch of the Farmington that one day hit the paydirt of the ancient central lowland and began to work northward so fast through the soft rocks that it captured several adjacent streams, diverting their waters into its own channel and so giving birth to our present Connecticut River.

When the glacier in time came down, although it left its mark everywhere throughout the length of the valley, it had little effect on the valley's main form. The ice moved smoothly down through the long trough that generally paralleled its own direction of movement, finding little resistance to its progress. Between Saint Johnsbury and Littleton, however, the valley makes a jog, so that for a little way it lies east and west, at right angles to the direction of the glacier's flow. This crosswise bit of valley dragged at the underside of the passing ice and in time became clogged with bouldery rubble that remained to dam the river when the ice finally melted away. In the years since the ice has gone, the river has dug itself a new channel through the glacial debris; but borings show that this channel lies half a mile north of the old, preglacial one. In this reach the modern river flows through a mass of huge boulders from which all smaller rocks and soil have long since been washed away. Over this the water drops 320 feet in the course of some twenty miles. The roughness is now concealed by a man-made reservoir; but the place still goes by the name of Fifteen Mile Falls.

The 360-mile-long valley of the Connecticut is a composite structure, and its upper, middle, and lower part each has its own character and its own history. The upper valley has developed in a straightforward way by headward erosion through bedrock, following the course of least resistance. The headwaters region is still rough and steep; but below Colebrook it flows over relatively softer rock, and here the river has worn its bed low and smooth. Its tributaries, however, have had to work across the grain of the land. Erosion is much slower in this direction, and the side valleys still remain youthful in form, with steep sides and strongly sloping streambeds that are full of falls and rapids.

From Stewartstown down to Greenfield the river flows in a long, winding groove through hills that are covered now with northern forest, now with grassy pastures, but always intensely green. The valley floor is a flood plain, wide or narrow as topography and the hardness of its foundation determine. Over this the river meanders from side to side, eating into the confining valley sides in time of flood and building up the flat valley floor as the water subsides.

Here and there on the valley floor are little conical kame hills that were set down by the glacier, or the long winding ridges of eskers. Rising from the flood plain are rank on rank of terraces, steep-sided and level-topped. Highest of all rise the rolling hills and mountains of bedrock. All of the features that stand between valley bottom flood plain and upland hill-slope were constructed by meltwater from the glacier during its final dying days. Some of them are especially striking in the upper reaches of the valley.

The higher terraces, especially along tributary valleys, mark the locations of postglacial ice-margin lakes that were held against a hillside by a remnant of melting ice lying in a valley trough. Now that the ice side of the lake bed is gone, the old bottom deposits are left as terraces hanging on the hillsides. Each successive terrace down the valley side marks a halting point as the rotting ice mass and its imprisoned water fell to lower and lower levels.

Such "kame terraces" can be recognized by their faintly undulating tops, often marked with kettle depressions left where a buried ice block finally melted out after the lake had been drained. The sloping terrace sides that once lay in contact with the valley ice may show sharp pits and ridges that are impressions of the fretted lobes of wasting ice. Kame terraces bear no particular relation to the present streams at their feet and may face in any direction —upstream, downstream, or sidewise.

The lowest terraces are the remains of the floors of old Lake Hitchcock or Lake Upham. Lake Hitchcock * was formed early in the course of the glacier's last demise when a mass of rocky debris accumulating from the melting ice formed a dam across the valley near Middletown, Connecticut, just at the upper end of its rocky gorge section. The appearance of this blockade had far reaching effects. Water was impounded from there northward all the way to the retreating edge of the ice. The resulting lake eventually reached as far north as Lyme, New Hampshire, a distance of 157 miles. In Massachusetts and Connecticut the lake was as much as twelve miles wide; farther north it narrows to two or three miles.

The shoreline of Lake Hitchcock can be traced all the way around, clearly marked and in some places complete with beaches. The old lake floor is covered with varved clays that are as much as a hundred feet thick. From its well preserved shorelines and from its bottom deposits one can reconstruct the history of this extinct lake.

Counts of the clay varves reveal that Lake Hitchcock lasted for more than four thousand years. Careful studies of the old shoreline show that throughout that long period of time the water level remained remarkably constant. But at present the old waterline is distinctly higher in the north, standing 657 feet above sea level at Hanover, New Hampshire, and only 135 at Middletown. Since then as now the surface of a lake lay horizontal, such a sloping shoreline calls for an explanation.

The odd position of the shoreline is a result of the tremendous weight of the ice sheet. When the glacier was at its height, the crust of the earth actually sagged beneath the pressure of thousands of feet of ice, more so at the center where the ice was thickest, less so at the thin southern edge. It was while the earth was thus deformed that Lake Hitchcock had its existence. Then, as the ice

* Named for an early geology professor at Amherst College, who was one of the first to recognize the existence of this ancient extinct lake.

burden wasted away and its meltwater ran off into the sea, the land gradually rose again and returned to its original conformation, tilting the ancient shoreline upward at the north as it did so.

The same kind of upward rebound is still going on today in the far north around Hudson Bay and the Baltic Sea, where the ice sheet has disappeared even more recently. In these places the sea is becoming shallower, especially at the north; and in northern Scandinavia new land has appeared above the sea just within historical times. Marks placed on the rocky shore show that the land has risen as much as three feet in a century.

Lake Hitchcock came to an abrupt end when its water level suddenly dropped by a distance of ninety feet. The evidence suggests that this happened in a single year when water burst through the barrier below Middletown in the spring breakup and floods. The subsidence of the water level was enough to drain the valley immediately as far up as Charlestown, New Hampshire. North of this the land was still at that time depressed by the ice, so that the upper part of the valley remained filled with water, forming what is known as Lake Upham. This lake in turn grew northward against the retreating edge of the glacier, developing branches that reached up the side valleys to Littleton, New Hampshire, and Randolph, Vermont. In its later years, before the ice had completely left New England, Lake Upham was in turn fed by slightly higher tributary Lakes Coos and Colebrook.

In all of these lakes, sediments accumulated to depths of many feet. In the years since the lakes were drained, the river has carved its valley deep into the soft silts and clays of the old lake bottoms, leaving the remnants shelving up into high-standing terraces. Such river-cut terraces lack the pits and hummocks of ice-margin kame terraces. Instead their sloping sides show the smoothly rounded, bitelike marks made where wide-swinging meander loops of the river undercut the bank. As the meanders change their positions, the incised scarps remain, their concave sides always facing the present river. Sometimes the scalloping of the terrace edge is interrupted where an outcrop of rocky ledge has resisted the washing of the river. On terraces formed of remnants of the floor of Lake Hitchcock stand the New Hampshire towns of Walpole, Lebanon, and Hanover. Part of Hanover also spills down onto the bottom of Lake Upham.

Still another kind of terrace in the modern valley consists of deltas that were built where sediment-laden streams poured into the ancient lakes. These are made, not of clay and silt like the floors of quiet lakes, but of coarse sand and gravel dropped by fast-flowing river currents. Claremont and Plainfield, in New Hampshire, stand on large gravel terraces that originated as river deltas.

There is one remarkably long delta that was built by a river flowing south out of the edge of the glacier into the long valley lake. As the ice melted back, the end of the lake and with it the mouth of the river and the head of the delta moved northward too. Before Lake Upham disappeared, the delta had grown into a long, winding ridge or esker that reaches from Windsor, Vermont, to Lyme, New Hampshire. The gravelly ridge was later partly buried by clay deposits, but the modern river has since excavated enough of it so that it can be traced for miles along one side or the other of the valley.

Above the highest terraces rise the bedrock hills. Such soil as they have on their bare bones consists of glacial till or ground moraine, a heterogeneous mixture of boulders, sand, and clay.

From Greenfield to Middletown the hills fall back and the valley opens out into a broad, level lowland. The relatively young, soft rocks that underlie this section of the valley were laid down as sediments in Triassic times, some 180 million years ago. At that time the valley already existed as a trough between two rows of mountains. Then suddenly (as geological time goes) a hundred-mile-long crack tore through the crust of the earth just at the foot of the eastern range. Land lying west of the fissure line then sagged far downward, while the hills on the east reared up to form tremendous block mountains. In the course of time, mountain torrents racing down the steep eastern slopes filled the valley trough with sediments that reached a depth of several miles. Eventually the soft deposits of sand and silt were compacted into layered rocks that still bear imprints of events of those remote times.

All of this once spectacular topography was then completely erased by long aeons of erosion, and for a time most of New England lay as flat as a table top. In the past seventy million years, the planed surface has again been uplifted; and as vigorous erosion set in once more, differences in rock resistance imposed their pattern on the newly etched landscape. The softer Triassic sedimentary rocks have been rapidly worn down almost to sea level, leaving a wide lowland that is streaked with ridges of more resistant traprock. On east and west the lowland is flanked by rolling uplands that are made of hard, crystalline rocks like granite and gneiss.

This central section of the valley is a veritable geological museum. Among its most interesting exhibits are the outcroppings of rocks bearing dinosaur tracks that lie exposed at the surface in a number of places between Turner's Falls, Massachusetts, and Durham, Connecticut. The track-bearing rocks are layered sandstones. Some of them have been tilted and broken, but otherwise they have not been distorted in all those millions of years. Their flat expanses bear clear-cut three-toed footprints of all sizes and shapes. The dinosaurs that made them seem to have traveled in herds, since the tracks in one place all head in the same direction. The footprints meander over great areas of ripple marks that must have been made in quiet, shallow water. Some of the rippled surfaces show also a spatter of drop marks that look like the splash from the edge of a passing shower.

A curious observer is bound to wonder how a record of the dinosaur's casual wanderings and the watermarks of trivial, long-ago events came to be so clearly preserved. Perhaps the circumstances do not seem strange to those who know our western desert basins. Recall that in Triassic times this lowland valley was enclosed between high mountains on east and west. Down from these came large quantities of silt in the runoff water of every rainstorm. This spread out over the flat lowland in what must have been alternately wet, muddy flats and sunbaked deserts. The silty mud made a fine surface to receive sharply detailed imprints. When the mud had baked in the sun to a hard, dry surface, the marks in it were preserved against destruction by the next influx of mud and water. In our own West there are similar places, where cloudbursts in the mountains give rise on the flats below to temporary streams and ponds that soon dry up and disappear to leave bone-dry barrens. Since the dinosaurs have disappeared from the valley, the sediments bearing their tracks have been deeply buried, compressed into rock, lifted up into the air, and at last re-excavated by erosion.

Another exhibit in this valley museum are the well-known trap-rock or basalt ridges that rise boldly up from the valley floor. Such are the Hanging Hills of Meriden, East and West Rocks and the Sleeping Giant near New Haven, and many less well-known rocky eminences. These are all formed by the protruding edges of lava sheets that lie buried among the layers of softer rocks. Long ago when the valley floor sagged toward the east, its rocky basement became tilted. As erosion has etched away the weaker sedimentary matrix, the resistant margins of the harder lava layers are left standing out in sharp relief. The western sides of the larger lava outcrops form precipitous escarpments. From the bare, palisaded faces of these, solid rock splits off in vertical hexagonal columns that collect in broken fragments along the base to form coarse talus slopes.*

In central Massachusetts the lava sheet whose edge forms the Mount Tom and Holyoke ranges crosses the line of the rift that so deepened the primordial valley in Triassic times. As the broken rocks slipped along the fissure line, land on the eastern side moved southward at the same time that it rose; and in this way the present Holyoke Range was pulled round until it came to lie east and west, almost at right angles to the other trap ridges.

In this low-lying valley the 878-foot eminence of Mount Holyoke ** offers a view in all directions, as far out across the country

* The palisades of the Hudson and a number of ridges in northeastern New Jersey, including Watchung Mountain, were formed at the same time and in the same way.

** The summit is accessible by auto road.

as the condition of the atmosphere permits. As you stand on the rocky crest, the great Connecticut River lies directly below. Your eye can follow it as it comes out of the hills at the north, slips through the gap between Mount Holyoke and Mount Tom to the west, and glides on southward across an even wider lowland to Springfield and beyond.

Near the base of the mountain lies an oxbow lake that dates from about 1830. Just to the north of it are two sharp meander loops that will some day become oxbows in their turns. One loop is nibbling at the northern edge of the village of Hadley and at the same time adding to a peninsula that reaches east from Northampton. If the light is right, you can detect the curving swales where the river once flowed and where even now it washes in at flood season. Farther north, toward Sunderland, you can see that the river is bordered on either side by low, natural levees that separate the channel from the flood plain.

If flood stages of the river are hard to imagine, stop when you go down to the highway again and look for the signpost at the roadside a little way to the north that shows the high water levels of various recent floods. As you stand on the pavement, some of the marks are higher than your head. Now turn all around and see what a mighty lake this valley holds in floodtime.

While you are still on the mountain, look around you where the upland rises on all sides of the valley. If the day is exceptionally clear you can see the even skyline of the Berkshires to the west and the rolling upland, not quite so high, to the east. In the north the low peak of Mount Monadnock may be visible, protruding above the general upland surface, or in the east Mount Wachusett, a smaller monadnock near Worcester.

From the mountain top you may be able to see how clearly the valleys of smaller rivers flowing into the Connecticut show the influence of rock resistance on the course of valley making. The Deerfield, the Westfield, and the little Mill River all emerge from steep, narrow valleys in the western heights and then flow a short distance over the lowland before they join the Connecticut. In the ages since the flattened face of New England was last uplifted, the hard upland rocks have been no more than nicked, while the softer Triassic inlaid rocks have been worn to a flat lowland almost at sea level. Though hard upland and soft lowland surfaces are of the same chronological age, the valleys cut into them have reached entirely different stages of development.

Below Middletown the Triassic lowland veers slightly to the west toward New Haven. But the river suddenly departs from the wide lowland and plunges into a narrow gorge that runs a little east of south, cutting right through the hard rock upland. This is most improbable behavior for a river and cries out for an explanation.

The best one available seems to be that back in Cretaceous times, which followed soon after the Triassic, the sea encroached over the southeastern part of the then flattened New England, possibly reaching as far inland as western Massachusetts. Through the course of several million years sediments from the land were deposited in the shallow sea's edge. When the land later rose again and the sea withdrew, the bottom deposits appeared as a flat coastal plain sloping off to the southeast, with the main rivers of the time flowing downhill across it.

The Cretaceous deposits were soft and have long since worn away from New England, although they still appear on Long Island. As the newer rocks gradually disappeared, rivers that had developed on their surface continued in their southeastward courses, biting into the hard underlying bedrock and entrenching themselves in their old courses. At the same time, certain lesser streams flowing south over weaker zones in the primordial rock worked faster and were able to capture much of the old southeastward drainage system. The Housatonic, like the Connecticut, shows the interplay of the two sets of forces, flowing for some distance southward, then veering toward the east in its lower reaches.

The entry of the Connecticut into its steep-sided lowermost section is abrupt and dramatic. As you drive east along U. S. Highway 6-A just beyond Portland, Connecticut, pull over to the roadside where the view opens out on the right. Across the valley in front of you lies a long green rampart made by the hilly upland where it rises sharply from the Triassic lowland. In the edge of it, clear as a textbook diagram, is the notch where the river turns into the upper end of its gorgelike lowest reach.

For the twenty miles from here to Long Island Sound the valley has its own special character. Its high, rocky walls are never far apart. In some places the river sweeps close against the foot of the valley sides. Small tributary streams enter through their own steep little gorges. Occasionally a wide place opens out where one of the larger branches joins the main valley. In these alcoves and elsewhere against the valley's old rock walls appear ice-margin kame terraces standing at various heights. At Chester and Hadlyme they rise one above the other like flights of irregular, free-form stairs. On such terraces up and down this reach of the valley stand the little towns that had their heyday a century to a century and a half ago.

As the great valley today bears the marks of the river's long history, so it shows also the marks of man's brief influence. The headwaters region has still an air of remoteness, although the first settlers trickled in there soon after the Revolution, when Rogers' Rangers had virtually exterminated the Saint Francis Indians and so made the country safe for white men.

Topography in this northern end of the valley is not such as to encourage farming enterprise, even in earlier days. From the river's source in the Fourth Connecticut Lake to Stewartstown, at the corner where New Hampshire, Vermont, and Quebec come together, is a journey of some forty-five miles and in this distance the valley drops from 2,625 to 1,020 feet above sea level. This is a rough and forested land that seems destined to continue indefinitely with a sparse population. People here earn their living from the pulp and paper industry and alternatively cater to the needs and desires of hunters and fishermen. Even today, almost two centuries after the first settlers, this region has little of the traditional snug New England look.

From Stewartstown down to Greenfield the river flows through an agricultural country, with the marks of the dairy cow visible everywhere. Settlement of this part of the valley was delayed by Indian warfare until the 1750's; but then it was rapidly occupied all the way up to Canada, and one after another in quick succession the lovely little towns sprang up that still have their white houses and steepled churches set about the elm-shaded village greens.

Pioneer settlers in this region began by selling the choicest timber from their land for lumber. The rest of the trees they disposed of by burning just to get them out of the way, or perhaps to make a cash crop of potash. Agricultural produce grown on the land thus cleared moved to its markets by way of flatboats on the river. The round trip between Wells River, Vermont, and Hartford could be made in twenty-five days, sometimes less. Navigation, however, was blocked by rapids at Bellows Falls, South Hadley, and Enfield. The first dams and canals at those places were built to carry flatboats around the rough water.

It soon occurred to enterprising Yankees that falling water could be used to turn machinery and soon new dams and factories began to make their appearance. The little towns that grew up around the mills have a core of gray, grim ugliness that is still surrounded by pleasant residential areas that speak of prosperous years. Older houses date from a period of architectural grace and simplicity. Towns that flourished chiefly in the post-Civil War era have their quota of Victorian exuberance that has not yet been overgrown by a later era of expansion.

At Greenfield the river emerges from the hills into the broad Triassic lowland. The valley below this is a long-inhabited country. It was in 1635 that the first group of settlers came overland from Newton, Massachusetts, to the rich meadows at Hartford. Within two years all the meadowland from there to Springfield had been occupied, and by 1700 settlement was continuous between Greenfield and New Haven. This was America's first inland frontier, and the fact is commemorated in its adopted name of Pioneer Valley.

Expansion into the narrower valley above Greenfield was held in check for decades by warring Indians who repeatedly swooped down from the hills on the outlying settlements. In the dead of winter in 1704 the well-established little village of Deerfield was attacked and burned in a horrendous Indian raid. Many of its inhabitants were massacred outright, and most of the rest made a harrowing forced march through the snow to Canada, where some of the survivors lived out years of captivity. For another fifty years settlement ebbed and flowed from the frontier forts that were maintained at Brattleboro, Walpole, and Charlestown.

Meanwhile the middle valley prospered, as witnessed by the many fine old houses still standing in its towns and countryside. The silts and clays of old Lake Hitchcock in this region make some of New England's most fertile agricultural soil, and certainly its most abundant. The best soils are used for growing tobacco, onions, asparagus and other truck crops. Dairy farms are common on moderately good areas. The poorest coarse sand to this day is covered with pitch pine forests.

Hartford lies in the middle of this flourishing farm country. It is also located strategically at the head of navigation on the Connecticut. In the days of small sailing ships and bad roads Hartford and Middletown were important and thriving seaports.

The lowest part of the Connecticut valley, like the Maine coast, has many sheltered coves where deep water is rimmed with flat-lying land, however narrow the rim may be. The nearby forests were originally full of excellent oak timber. The natural outcome of this juxtaposition was a great development of ship building and waterborne trade, both with the local hinterland and nearby coastal towns and with lands halfway around the globe. Steam and iron and railroads eventually killed all this, and nothing much has happened to change the villages for a hundred years. Now they have been taken over by summer people and others who have city resources to draw upon, and most of them are beautiful but sleepy places. Their present-day life revolves around leisure and recreation, and every cove is full of pleasure boats of all sizes and descriptions.

The lowest end of the valley is an estuary, influenced by the rise and fall of ocean tides. Near Hadlyme the shores begin to show this, and from Essex on, the river is flanked by broad expanses of tidal marsh. Here one begins to sense the flavor of shore resorts, with beaches and bathers and saltwater fishermen, both sporting and commercial.

Even to the end the valley bears the mark of the glacier. The last flat-lying arms of land that reach out around the river's mouth are the remains of old ice-margin terraces, formed when Long Island Sound was a valley filled with the last stagnant remnant of the continental ice sheet. Even the headlands and islands of the Sound are the work of the dying glacier. Only when it sweeps eastward past New London into the rough currents between Montauk Point and Block Island does the water from the heart of New England finally escape to the ocean.

12

New England's Farmlands

TWO HUNDRED YEARS ago the vast majority of New Englanders got their living from the land. Nearly all the settled area lay in open fields and pastures, and the forest began at the frontier. Today three-quarters of New England is covered with woods, and the rest bears one of the densest populations in the United States. Yet in 1950, of the 2,500 counties in the nation, Aroostook County in Maine ranked sixteenth and Hartford County, Connecticut, twenty-sixth in the total value of their farm products. Behind these bald facts lie three full centuries of agricultural change and development.

From the very beginning enterprising and ambitious Yankee farmers tried hard to find a staple cash crop that could be grown in large quantities on a plantation system such as the southern colonies used, but the lay of the land defeated them. There just are no large expanses of land with uniform soil and moderate topography where cheap but inefficient slave labor could be used profitably. So the farms always remained small, family-size affairs.

By the time of the Revolution, however, an economic differentiation had set in, and New England farms generally began to fall into two rather distinct groups. Those located on fertile, relatively level land were nearly all commercially profitable from the start, and when the growth of nearby towns provided a reliable market for cash crops, they prospered mightily. Many such farms, along with the large, handsome houses built with their proceeds, are still in use today. Some of the substantial rural houses dating from the late 1700's or early 1800's stand on smaller farms that flourished while the nation was young, but did not survive as working farms because they were not able to adopt the large-scale operating methods that later became necessary to meet competition from the developing West.

In lowland New England more old houses than one might think represent wealth from ship-building and trade. Many a minor inlet from the Connecticut River or Narragansett Bay that now harbors only pleasure craft was an important shipping point in the days when roads were seas of mud in wet weather and dust in dry, and when land freight moved according to the strength and speed of oxen. But even discounting the influence of maritime trade, there are still many fine old houses that clearly had their origin on the land.

In the steeper, stonier hill country, on the other hand, much of the soil was too poor, or often merely too scarce, for profitable farming even when the land was first cleared. There the farms are smaller, and houses and barns are far more modest in both size and style. Many of them were occupied for only a generation or two and then deserted, left to a fate that has obliterated them except for cellar holes and stone walls running through the second-growth woods.

Many a small farmstead that was marginal or worse from its first days has been preserved, not with the proceeds of agriculture, but only because the people who lived there could find work in the vicinity for wages that would piece out their inadequate farm income. Some of our earliest part-time farmers probably worked in the little water-power factories scattered along small, fast rivers in the rural hinterland—some of them still in operation and stirring wonder in the visitor from the newer West.

The years around the time of the Civil War were a major turning point in the history of New England agriculture for several reasons.

The decades just before that saw the development of large agricultural machinery that made small-scale, self-sufficient farming no longer competitively profitable. By the 1860's transportation from the fertile, level West had been so much improved that the small eastern farm was no longer really needed as a source of food for the nation's people. And with the rapidly increasing demand for industrial labor, farming was no longer practically the only way for an ordinary man to make a living. The change began as early as the 1830's and gathered momentum rapidly. By the 1860's more and more farmers were pouring down out of the hills, leaving their stony fields to grow up to brush or field pine.

But, even as the flood of abandonment went on, some people were finding new ways of doing things that made it possible to live on the farm and still make a decent living. One important change has been a steady increase in part-time farming. In recent years more than a third of New England farmers have worked an appreciable part of their time off the farm. Often the outside job is in some manufacturing industry, or it may be seasonal work as a carpenter or mason or driving a school bus. Many of the older farms that have large, substantial buildings but an acreage that is no longer large enough to yield a good income are operated successfully on a part-time basis. And of course large numbers of the handsomer old houses are used as country homes, either for summer or year-round use, and the places not farmed at all.

New England has shared in the modern trend toward larger farms. Although land continues to pass out of agricultural use, much of it is clearly not suited to farming under present-day conditions. The acreage that is still farmed is used more intensively, a greater proportion of it being improved rather than left, for example, as rough pasture. At the same time the size of individual full-time or commercial farms is increasing, although most of them still remain family-size enterprises. This increase in farm size is accomplished by consolidation of old land-holdings, two or three old places being run as one farm, or by buying or renting land from part-time-farmer neighbors or from city people who hanker to live in the country on a real farm but have no taste or time for the life of a real dirt farmer.

Farming a larger acreage with no increase in manpower calls for mechanization. This offers difficulties in a country that is divided into small fields by solidly built stone walls, with the fields often strewn with rocks besides. A modern bulldozer can easily cope with these, however, and increasingly, old walls and surface rocks are being removed, or if there is no other way to dispose of them, piled in a corner or even buried. The enlarged fields can then be contour-cultivated to control soil erosion and tended with tractor-drawn machinery. Manufacturers have responded to this development by making smaller tractors for use on the smaller eastern farms.

Along with the trends toward part-time farming and toward larger farms has gone an increase in specialization. Since no point in New England is more than an overnight distance from large cities with their teeming millions of consumers, perishable and luxury products have a good market readily available. Although crop rotation requires a minimum degree of diversification, the modern farm here is basically a one-crop establishment. New England's rural scenery varies widely according to the local specialty. The informed traveler finds much to interest him in what he can see as he goes about the countryside.

Dairy farms are a general regional specialty that could be likened to a matrix in which all the others are embedded like pieces of enamel. They range from small, not very prosperous places in the back country where half a dozen scrawny cows scrape a living among the spruce or cedar, to huge establishments with generous, well-kept barns and silos and lush green fields where large herds of sleek cattle find life very easy and can devote all their energies to producing milk. Perhaps the extreme is the estate of the gentleman farmer who can offer his guests a choice of milk or champagne as equally costly delicacies.

The raising of cattle as a cash crop began long ago when self-sufficient subsistence farming was the source of most people's living. A large proportion of the land is too steep and stony for any agricultural use except rough grazing. When coastal towns began to develop as thriving maritime centers, it was natural for the hinterland to turn to producing food for the town population, especially food of a kind that could get to market on its own legs. One of our first fairly passable roads was the Pilgrim Path, which provided a route for cattle raised in the relatively fertile country around Plymouth to get to Boston and the other growing towns around it.

For a long time cattle were used chiefly as a source of beef and hides, and milk was sold in the form of cheese or at least butter. But since roughly the turn of the last century fast railroad transport and especially adequate refrigeration have made it possible to bring large quantities of fresh fluid milk to the cities. Most of New England is divided between the Boston and the New York "milk sheds," the drainage pattern being determined by the form of the railroad network. In the case of Vermont, for instance (which has about as many cows as people), milk from the western side of the state appears presently on the doorsteps of New York, and milk from the eastern side goes to Boston. The divide appears to bear some relation to the language divide between more or less standard American on the west and a rich Yankee twang on the east. Some milk is still made into butter and cheese and, of course, ice cream, but farmers get a better price for the same milk if it is destined for the milk bottle or the cream pitcher. So most of the cheese now eaten in New England comes from dairylands a few days' travel farther west, and the product of local cows is used as fresh milk.

Like everything else in modern life, dairy farming is becoming mechanized, from milking machines and coolers in the barn to haywagons in the field. The horse long ago gave way to the truck and tractor, and the pitchfork is rapidly disappearing before the spread of the mechanical hay baler. Even the hay is often dried mechanically in the barn, and our grandchildren may one day ask, "Grandpa, what does 'Make hay while the sun shines' mean?" But let the sentimentalist who is willing to walk five blocks from his parked car and run his office without any business machines, even typewriters, cast the first stone.

Not only the tools that are used, but even the land itself shows the effect of changing times. The rough hill pasture, strewn with fern-girt boulders and threatened by invading trees, is a familiar and picturesque part of the New England landscape; but its days are surely numbered. As long ago as the eighteenth century the more enterprising farmers began to seed their pastures with good forage plants such as clover and timothy and to fertilize with manure and lime. Even a good pasture, if it is neglected, gradually deteriorates, and in a few years the field will support very few grazing cattle and will not be worth the trouble to mow it for hay.

In present years farmers are making less use of permanent pasture of any kind. The more promising fields can be cleared of stones with a bulldozer; the impossible ones are better left to revert to woods. The improved fields are then placed on a rotation system that includes several years of a hay or forage crop and then one of a plowed row crop such as corn. The periodic plowing turns the grassy turf under where it will decay to enrich the soil and especially to improve its texture. At the same time fertilizer can be plowed in and then a new seed mixture planted. In this way a farmer can have vegetation with any combination of characteristics he may desire. Even in this climate, one of the stock grower's serious problems is likely to be inadequate pasturage during the warm, dry weeks of summer, and it is here that plowed and seeded fields are so superior to any kind of permanent pasture.

Other dairy farm improvements are ditching and draining of the wet spots that may appear anywhere in this glaciated land, and the construction of watering ponds at strategic points. Ponds can serve not only to govern the goings and comings of animals, but if they are large enough and accessible to buildings, they may greatly reduce the cost of fire insurance and encourage a farmer to carry this kind of protection for the large investment that a modern farm represents.

Next to cattle and running a close second, poultry farming is the most important kind of agricultural enterprise, in terms of both the number of establishments and the amount of income they produce; and the number of chickens that live in modern apartment-style henneries throughout New England must be astronomical. Here again the nearness of mass markets makes all the difference. Many a farm that was submarginal for anything else has become a successful poultry business, whether on a full-time or part-time basis. Hen houses large enough for thousands of birds occupy little area. The nature of the land makes no difference whatsoever, since well-reared chickens and turkeys ordinarily never set foot on the open ground; and feed grains are mostly bought, although whether they are produced at home or imported from the west depends on relative costs at the time.

Besides cattle and poultry, New England is the home of the McIntosh apple. Peaches are grown in a few especially favorable places. Truck crops are raised in any quantity only where fertile soil lies near the larger cities. Although roadside stands in the middle Connecticut Valley offer a temptingly wide variety of local produce, the bulk of fresh fruit and vegetables eaten by the large city populations comes from the South or the West.

One of the most intensively cultivated of New England's crops, with high production costs and high market value, is tobacco. Although most outlanders seem to believe that all tobacco comes from North Carolina, certain special kinds of it are grown extensively on fertile, sandy loam along the Connecticut River in Massachusetts and Connecticut. It has been cultivated as a cash crop hereabouts ever since colonial times, so much so that the early fathers tried to legislate farmers into using their land and their labor for more utilitarian crops. The town of Warehouse Point north of Hartford takes its name from a tobacco warehouse that was built there in 1825.

Tobacco land can be recognized by its broad, level, silty fields dotted with barnlike curing sheds. In the summer a traveler coming down from the hills has glimpses of large white rectangles lying here and there on the valley floor. Coming closer, he sees that the white patches are flat-topped tents, nine feet high and some of them covering many acres. If the tent sides happen to be up at the time, he will perhaps see a solitary man riding along on a small tractor in the light shade of the thin cloth roof, working back and forth with a cultivator. In early summer when the plants are still small, there may be a long line of hand-hoers moving slowly down the parallel rows of plants.

Spotted over the fields are tobacco barns, long, rectangular, pitch-roofed buildings the equivalent of two stories high. A few are painted barn red or white, but most of them show the silvery gray-brown of weathered, unpainted wood. The sides are made of vertical boards, every second or third one attached by hinges so that it can be swung open for ventilation during the curing process.

Connecticut Valley tobacco is used to make the outer parts of cigars. Leaves grown in the shade of cloth tents are very smooth and thin, with small veins, and are used for the outermost wrapper of the best brands. Open-grown leaves of the same varieties are also large and relatively thin and are used chiefly for the "binder" of cigars—a layer rolled around the innermost "filler" and in turn enclosed in an outer wrapper leaf.

Only a high value crop will repay the large amount of labor that is required for raising tobacco in this region. Moreover, the large tents as well as the tall, succulent plants are very susceptible to damage by wind and hailstorms. Bringing a crop successfully to harvest is always something of a gamble, and the decision whether or not to carry expensive crop insurance is a weighty one for a tobacco farmer.

Work on a tobacco farm begins in early April, when seeds are planted in coldframes. Skilled workers are needed to carry out all the procedures involved in soil sterilization, watering, ventilating, and insect and disease control under the glass sash.

Meanwhile there are the tents to take care of. These are put up every spring and taken down again every fall. They are made of a special kind of strong white cheesecloth, supported by wires that are in turn held in place by tall posts. The tops of the tents must be replaced with new cloth every year, although the old tops can be used again the next year for sides.

The young plants are taken up from the coldframes and transplanted to the fields in late April if they are to be grown under cloth, in early June if they go in the open field. Planting is mechanized as fully as possible. One man drives a tractor that pulls the "transplanter." This machine makes a pair of furrows and squirts a half-pint of water in the place where each plant is to go. A pair of "droppers" ride seated low down at the back of the transplanter. The dropper's job is to take the little plants from the bucket of water beside him and drop one on each damp spot in just the right way at just the right time so that the machine can draw the soil up over its roots.

As soon as the young plants have become established, machine cultivation begins. This is done every week through the summer. While the plants are still small they are also hoed by hand in order to loosen the soil and remove weeds within the rows.

By late summer the plants are pushing against the tent roofs. Now hand labor is required again for "topping" and "suckering"—removing flower buds from the tops of the plants and any branch shoots or suckers that may appear along the stems. This encourages the leaves to grow to their maximum size. At harvest time the leaves are broken off individually from the stems of shade-grown plants. In the open-grown crop the whole plant is cut off at the base. Then the crop is sewed by hand to special laths and hung top down from the rafters of the sheds to cure.

For a month or more the passer-by can see in through the ventilating slits or the large doors at the ends of the barns and watch the progress of the change from green through yellow to brown. All during this time conditions of temperature, humidity and air movement are carefully controlled by opening or closing the sheds and at times "firing" with many small charcoal or gas burners.

As soon as the fields are cleared, the tents are taken down and a quick-growing cover crop is sown, usually something grassy. This serves to prevent washing and blowing of the soil during the winter, although tobacco fields are ordinarily so flat that erosion is not a serious problem. More important is the fact that the cover crop absorbs what is left of the large amount of fertilizer used on the tobacco crop. In this way, instead of being washed out of the light soil by fall and winter rains and so lost, mineral nutrients are held in the grass tissues and returned to the soil when the cover crop is plowed under the next spring. This serves also to keep up the humus content of the soil.

Tobacco is grown on the same land continuously, with no rotation of crops. Oddly enough, the yield of tobacco is lower following some other crop, apparently because of an increase in the prevalence of a root-rot disease.

Parts of the tobacco region where the soil is slightly heavier are planted to onions and asparagus, the asparagus crop being on the increase at present. Onion growing makes use of the local population of farm women from central Europe who are willing to do the necessary "stoop labor" for thinning and weeding the minute onion seedlings. The summer plumes of asparagus and the rich blue-green of onion shoots make pleasing scenes; and words are a pale substitute for experiencing the fragrance of a large onion field in the warm June sun that fills this sheltered valley.

Another region of intensely specialized farming is the potato country of Aroostook County in northeastern Maine. This is the most recently settled part of the east. Towns are far apart, for New England, and the farmhouses are not very old. The rolling fields, carefully machine-cultivated, are set about with dark, wild-looking forests. Distant views show blue mountains rising at the horizon. It is not a snug and cosy landscape and looks hardly at all like older parts of New England.

When Henry Thoreau came here on a canoe trip around 1850 this was frontier country. All of central Maine was then undergoing its first extensive logging, and Bangor was the lumber capital of the nation. Farming settlers coming in at that time thought that their staple crop would be wheat; but they soon found that the particular combination of soil and climate here would produce whopping big crops of high-quality potatoes. Soon the acreage planted to potatoes began to rise rapidly. From the 1890's to the present it has steadily skyrocketed.

Maine potato land is rolling and hilly enough to warrant more soil conservation measures than are generally practiced. Crops are rotated to provide a periodic "green manure" that will keep up the humus content of the soil, although organic matter decays slowly in the cool climate and humus is not rapidly depleted. Soil has to be managed rather carefully so that its acidity is kept low enough to allow clover in the rotation scheme but high enough to keep down potato scab infection. Here, too, farming is highly mechanized, although no one has yet devised a harvesting machine that can distinguish between potatoes and the small stones that abound in these fields.

The cool weather here acts as a curb on pests, especially at higher elevations. By rotating crops and taking various other precautions, it is possible to produce completely disease-free tubers; and Maine as well as northern New Hampshire grow large crops of certified seed potatoes.

Every farm has its special potato barn, separate from barns for other purposes. Potatoes will not stand freezing, so the storage barns are built low and set into the side of a hill or banked well up the sides with earth for insulation against the early frosts and severe winters. Their chimneys indicate the need of artificial heat at times.

Potato growing is a big business in Maine. This is apparent from the size of the warehouses that line the railroad tracks in all the little towns. Roadside billboards advertise fertilizers and pesticides as well as special farm machinery. An uninformed traveler is quite unprepared for the modern, citified look of the little commercial hotel in a northern country town like Houlton.

Maine blueberries have been canned commercially since 1870. These are the wild, lowbush kind with sweet, small fruit. Highbush varieties that are grown under cultivation farther south are not hardy enough for this northern climate. The fruit comes from large tracts of blueberry barrens that lie along the coast eastward from the Penobscot River. There are similar places in some of the hilly parts of New Hampshire. All of them originated as dense growths of wild plants that developed on abandoned pastures and burnt-over timberlands. The soil they grow on is quite poor, and no more profitable use for it has been found to drive out the wild berry bushes.

About the only regular care given most blueberry fields is burning them over every two or three years. This serves to prune the bushes, which stimulates new growth and induces a heavy crop of fruit. If the plants are not burned, they soon become crowded and spindling, and their yield is drastically reduced. Burning is done in early spring when the soil is wet or even frozen, so that the roots and the mat of organic matter they grow in are not damaged. Some sort of fuel is needed for the process. Sometimes a light scattering of hay or straw is used, or a special type of oil burner like a flame thrower is drawn over the fields.

While the plants are in bloom hives of honey bees may be set in the fields to increase pollination and ensure a heavy set of fruit. After the bees are removed, plants may be dusted to control insect pests. Growers may also have to contend with such troublesome creatures as browsing deer and even such exotic notes as porcupines and seagulls.

Blueberries are harvested by hand with metal scoop-rakes something like those used for cranberries. The fields are marked off into straight lanes with string so that they will be picked over systematically. This also discourages pickers from skipping the plants that have fewer berries. Then the fruit is run through a winnowing machine to remove bits of stem and leaves, and packed into baskets or crates, all without ever being touched by human hands. A large proportion of the crop is canned, some of it is frozen, and a little is sold as fresh fruit.

Another crop native to New England is the cranberry that grows wild in wet sandy or boggy places where the soil is acid. Cultivated cranberry bogs appear as smooth, rectangular carpets of a glossy, dark green sunk into the surface of a rough, scrubby landscape. More than half the cranberries grown in the entire world are grown on Cape Cod and the adjoining mainland in southeastern Massachusetts. They are also grown in similar boggy parts of Wisconsin and southern New Jersey, but Massachusetts enjoys the advantages of less severe winters than the former and fewer pests than the latter. Besides, endless quantities of sand for mulching are available just over the drainage ditch anywhere in that part of New England.

A good cranberry bog requires first of all an acid soil, peaty or sandy, and an abundance of water that can be flooded on or drained away rapidly and on short notice. For starting a new cranberry bed one hunts out a nice wet maple or white cedar swamp. If it has a thick peaty soil the grower will largely be spared the trouble of adding fertilizer. The first job is to remove all the existing vegetation, cutting off the tops and then grubbing out roots and stumps and leveling the soil surface.

Next comes the water system. This consists of a storage reservoir and a system of ditches, probably with dams and pumps, by means of which to regulate the water so that it will stand a constant ten or twelve inches below the bog surface. There will be one ditch all the way around the bog and as many cross ditches as are needed to let the water in or out in a hurry over the entire area.

When the engineering construction is finished, the bed is covered with a two-inch layer of coarse sand. It is then ready for the plants. These are bought by the bushel from another grower, who mows them from his own bog with a scythe. They are trimmed into suitably-shaped cuttings and stuck in the sand eight to twelve inches apart each way. Then the water is let in until the sand is thoroughly wet, and promptly drained off to prevent waterlogging. In a few weeks the cuttings will root, and the plantation is under way. The first moderate crop will probably come in the fourth year. After that, with reasonable care and decent weather, one can expect a good crop each year for a century.

A good part of the care of a bog is concerned with regulating its water supply. The aim is to provide plenty of moisture but at the same time adequate drainage during the growing season. Complete flooding is done for protection from frost, in spring for the flower buds, in autumn for the berries, and to kill pests at other strategic times. Bogs are left flooded over winter to prevent winterkilling, which is caused not so much by cold as by drying. The vines are evergreen, and drying winds blowing over the foliage while the roots are unable to absorb moisture from frozen soil can be very hard on the tops. Roots will survive, though, and plants usually recover from winterkilling in a year.

Flooding or "flowage" for frost protection during the growing season is a tricky business. It must be done rapidly when it is needed, but the water must stay on no longer than necessary, so that the active plants will not be injured by lack of air. The radio frost warning service of recent years is a great help to cranberry growers in deciding just when to flood their bogs.

Cranberries flower in June and early July and the fruit ripens in September and October. The crop is harvested by "raking" by hand with a large, toothed wooden scoop. Repeated combing through the waxy leaves gives the wood a fine, soft lustre that is the envy of a furniture finisher. Power harvesters are available but are not very widely used. If seasonal labor becomes very scarce this will no doubt change.

Another native crop that makes its mark on the New England landscape is the sap of the maple tree. A third of the United States maple crop comes from the one state of Vermont. In fact, Vermont has become so firmly associated in the public's mind with maple products that a true Vermont Yankee is skeptical of the notion that a good deal of high-grade syrup comes from Ohio and even Michigan and Wisconsin.

The Indians knew about making maple syrup. For years after the country was settled, "tree sugar" and honey were the only available sweetenings in the back country, and white sugar was an expensive luxury. Now the tables are turned, and maple sugar is the luxury sweet, but so relished for its own distinctive flavor that the public readily consumes all the maple products that are made.

Practically all sugar bushes—or groves or orchards—are remnants of the original northern hardwood forest, and many of them have been in production for a long time. It takes some forty years to grow a tree large enough to be worth tapping. When the trunk is ten to twelve inches thick it will keep one bucket busy, and more of them as it grows. A trunk of twenty-five inches or more can take as many as four buckets.

A well-managed grove contains nothing but maple trees, and it is not used as a cattle pasture in summer. Somewhere in the grove there is a plain, weatherbeaten sugar house where the sap is boiled down into syrup. There must be fuel nearby, preferably from surplus young trees in the grove itself. Since it takes a cord of wood to boil the 875 gallons of sap needed to make twenty-five gallons of syrup, a big woodpile is accumulated during the winter. A sap storage tank is located in a cool, shaded spot where it can easily be filled from the sled tank and where sap can run from it into the evaporators by gravity.

Along in March as the midday sun recovers from its winter enfeeblement people in the north country begin to think about the sugaring. Sugar weather is often the kind when vapors from melting snow hang in a murky haze over the soggy land all day, and only in the chill of evening does the air become clear and crisp again. There is a certain kind of heavy, wet snow with large flakes that stick together in clumps that is known in Vermont as "sugar snow."

Now when nights are still cold but noondays are pleasantly mild, the sap begins to run. All during the preceding summer the trees' green leaves went about their business of photosynthesis, making sugar out of air and water and sunshine. A large crown bearing a multitude of leaves produces more sugar than is needed by the tree for either current consumption or new growth, and the surplus is stored away in root and trunk. It is this surplus that begins to stir as spring approaches.

Just what makes the sap move we do not know. We do know that living tissues are necessary and that sap runs out of the tapped trunk in response to changes in pressure within. Pressure changes are in turn brought about by alternating high and low temperatures. When the day remains cold or the night warm, sap will not run. The year's sugaring is over for good when leaf buds begin to swell.

Maple sap is collected drop by drop in buckets that hang from the tapping spout or spile, a tube inserted into a small hole drilled about two inches deep into the trunk of the tree. Modern buckets are fitted with lids designed to shed rain and keep out miscellaneous debris. Still newer are flexible plastic bags. The sap must be boiled as soon as possible, before it can ferment or turn sour. So on days when the sap is running in a steady drip, there is a constant round of the grove with the collecting tank, and fires burn under the big, flat evaporator pans far into the evening.

In a few groves with closely spaced trees and an appropriate lay of the land, a system of pipelines is used to carry sap directly from trees to sugar house. The pipes are usually made of tin tubing and are suspended above the ground by wires attached to tree trunks and posts. This system saves labor during the season, but it is not without its special maintenance problems, although the new plastic tubing may simplify matters. Nevertheless, it will probably be a long time before the ox- or horse-drawn sled tank disappears from the sugar bush.

Not many farmers regard the product of their woodlots as a cultivated crop. There are occasional Christmas tree plantings and a few timber growing places that call themselves "tree farms." This is an ear-catching term intended to point out that the forest is being managed for continuous production and cared for and harvested as carefully as any other crop. But the ordinary farm woodlot is another story.

Foresters point out that the sad state of most woodlots is quite unnecessary, as we already know enough to be able to convert even a much abused red maple swamp to growing hardwood saw timber at a profit. The problem appears to be mostly one of education and persuasion, to convince the small holder that it is economically both feasible and profitable to manage his little acreage on good sylvicultural principles.

Tree planting is rarely necessary in New England unless one is starting on cleared land such as an old field or pasture. The work to be done in an existing woodlot consists chiefly of weeding out inferior species and individuals and thinning and perhaps pruning the crop trees. Harvesting begins with small fuel wood and progresses through poles and posts to the real cash-producer, sawlogs.

Perhaps the reader would like to know how to recognize a good small forest when he sees one. In the first place, both fires and cattle are strictly excluded. The ground is covered with a litter of twigs and leaves, with a good layer of decomposing humus beneath it. There are no over-ripe or weed trees, and no diseased, damaged, crooked, or "wolf" trees. The last is an individual that spreads its own crown so widely that it suppresses its neighbors over a considerable area. Every tree not destined for the harvest has a function either as a future replacement–advance growth–or else as a trainer to inhibit side branches on the larger trees. Good trees may even be pruned of their lower branches in order to prevent the formation of knots and eventually to produce one or two log lengths of solid, clear wood. Leafy crowns are well developed along the top third of the larger trees. In an even-aged stand the crowns almost touch to form a closed canopy, with nothing growing beneath them. An all-aged stand forms no distinct top and has many small seedlings and saplings coming along in the undergrowth.

The woodland produced by intensive sylviculture is a pleasant as well as a productive place. It is no spiritual substitute for a truly natural, wild forest, but it makes a handsome addition to New England's array of cropland scenery.

* * *

In the three centuries of its history, New England agriculture has gone through successive periods of expansion, retrenchment, and reorganization. One well might wonder what the future will bring. It seems likely that present trends will continue for some time—fewer but larger farms more intensively cultivated, and the roughest land reverting to forests that will be managed with at least the rudiments of sylvicultural methods. Suburbs will continue to spread, especially along the New York-Springfield-Boston corridor. But it is hard to see how even the bulldozer and the corporation farm can completely spoil this land whose character shows through in its good granite bony structure covered sparsely with a sinewy flesh that was worked over so thoroughly by the glacier.

Events in New England's Geological History

Index

Events in New England's Geological History

(For actual sequence read back from the end)

Era	*Period*	*Events*	*Years Ago*
CENOZOIC	Pleistocene	Climatic optimum: Time of maximum warmth	6,000-4,000
		Ice left northern New England	12,000
		Last retreat of ice began	15,000
		Glaciation began	1,000,000
	Tertiary	Renewed uplift brought on erosion of new steep valleys in old broad valley floors	12,000,000
		Broad, gentle uplift brought on renewed erosion. Valleys developed to broad open form. Cretaceous deposits eroded away. Soft Triassic rocks eroded to form central lowland	60,000,000
	LARAMIDE REVOLUTION. Birth of the Rocky Mountains		70,000,000
MESOZOIC	Cretaceous and Jurassic	Coast subsided and sediments deposited on its edge. Rivers flowed toward southeast	150,000,000
		Erosion of entire region to form peneplain with monadnocks. Original surface of present New England upland formed	
	Triassic	Sediments from nearby mountains accumulated in central valley, interbedded with lava sheets from volcanoes. Dinosaur tracks imprinted	
		Central valley formed by 100-mile rift in rocks and subsequent movement of these	180,000,000

APPALACHIAN REVOLUTION. Birth of Appalachian Mountains. General uplift of land surface with cracking and folding of rocks — 200,000,000

Era	Period	Events	Years ago
PALEOZOIC	Carboniferous	Coastal margin sank. Vegetation deposited in swamps around Boston and Providence that later made coal	250,000,000
	Devonian and Silurian	Folding and cracking of land surface. Ancestors of White Mountains formed by intrusion of granite from below	320,000,000
	Ordovician	Islands rose further and land area increased, especially to the east. Limestone deposited in western New England. Folding gave rise to Taconic Range	450,000,000
	Cambrian	Ancestors of Berkshires and Green Mountains rose as a chain of islands in an inland sea	500,000,000

Index